SpringerBriefs in Water Science and Technology

SpringerBriefs in Water Science and Technology present concise summaries of cutting-edge research and practical applications. The series focuses on interdisciplinary research bridging between science, engineering applications and management aspects of water. Featuring compact volumes of 50 to 125 pages (approx. 20,000–70,000 words), the series covers a wide range of content from professional to academic such as:

- Literature reviews
- In-depth case studies
- Bridges between new research results
- Snapshots of hot and/or emerging topics

Topics covered are for example the movement, distribution and quality of freshwater; water resources; the quality and pollution of water and its influence on health; and the water industry including drinking water, wastewater, and desalination services and technologies.

Both solicited and unsolicited manuscripts are considered for publication in this series.

Ariel Dinar • Mehdi Nemati • Sravika Pillarisetty

Water Institutions and their Performance

A Comprehensive Review of the Literature

 Springer

Ariel Dinar 
School of Public Policy
University of California, Riverside
Riverside, CA, USA

Mehdi Nemati
School of Public Policy
University of California, Riverside
Riverside, CA, USA

Sravika Pillarisetty
School of Public Policy
University of California, Riverside
Riverside, CA, USA

ISSN 2194-7244   ISSN 2194-7252   (electronic)
SpringerBriefs in Water Science and Technology
ISBN 978-3-032-24648-6   ISBN 978-3-032-24649-3   (eBook)
https://doi.org/10.1007/978-3-032-24649-3

We dedicate this book to our families, who always support us in our work and understand the importance of governance and organization in all life actions, especially where scarcity can easily add to chaos and jeopardize our management of limited resources.

Preface

Rising water scarcity and declining water quality across many regions of the world are prompting a wide range of policy interventions to address these challenges. Among these, institutional reforms play a central role in shaping policy responses in the water sector. However, water institutions are less tangible; thus, their importance in managing water resources is less appreciated. This review paper synthesizes works on water institutions worldwide across various sectors and over time, intending to identify commonalities across all sectors, locations, and time periods. The analysis uses a categorization system (bins) that enables the identification, contrast, and comparison of specific aspects of institutional analysis. Our review of 275 studies over three decades (1995–2025) suggests that water institutions are highly likely to address water resource management successfully during periods of scarcity. When the required conditions for their performance are unmet, the analyses showcase that institutions are prone to failure. One important finding from our study is that studies evaluating institutions and suggesting institutional development efforts are conducted in countries or river basins that are already strong and "institutionally savvy," rather than in countries that lack capacity, encouragement, and improvement of institutional development in their water sector.

Riverside, CA, USA	Ariel Dinar
Riverside, CA, USA	Mehdi Nemati
Riverside, CA, USA	Sravika Pillarisetty

Acknowledgments

Breanna Powell played a significant role in the early stages of data collection and dataset organization. Niloufar Nasrollahzadeh provided research support by initiating the population of the paper's sections with summaries of several of the reviewed papers. Rathinasamy Maria Saleth provided insights during the early stages of the work. The work on this paper benefited from the W5190 Multistate NIFA-USDA-funded project, "Management and Policy Challenges in a Water-Scarce World."

Competing Interests The authors have no competing interests to declare that are relevant to the content of this manuscript.

Contents

Chapter 1
Introduction

Abstract The introduction provides the motivation to the work, realizing that water availability and its quality have become limiting factors for economic growth of various nations and leading to inefficient use of water and loss of welfare. Increased impacts of various processes, including intensification of climate change impacts on various water-dependent sectors, population growth, and a rise in standards of living in different parts of the world calls for institutional interventions that were proven successful in recent years and various places. The monograph reviews the development of various institutions in the water sector, implemented under different contexts, time periods, scales, types of water use, and several other specifications. The Introduction describes also the organization of the monograph.

Water availability and its efficient use are influenced by several global phenomena (e.g., climate change) and local factors (e.g., widespread investments in water-related infrastructure and policies aimed at alleviating water scarcity). In recent decades, with the intensification of climate change impacts on various water-dependent sectors, population growth, and an increase in standards of living in different parts of the world, water availability and its quality have become limiting factors for the economic growth of various nations, leading to inefficient use of water and loss of welfare (Dolan et al. 2021). However, observing these trends across countries suggests that water scarcity does not necessarily decrease welfare everywhere. Even countries that face relatively high water scarcity can fare well because they have better-functioning water institutions (Araral 2010; Araral and Wang 2013; Özerol et al. 2018; Saleth and Dinar 2004).

The importance of water institutions in managing water resources under scarcity has been recognized in the economic literature (Jiménez et al. 2020; Kiparsky et al. 2017; Özerol et al. 2018; Saleth and Dinar 2004). Various studies have identified and quantified the important role that water institutions play in sustaining water resources, and subsequently, the rest of the economy. As survey results by Dinar and Tsur (2014) indicate, the 1980s and 1990s witnessed a surge in studies focused on the performance of water institutions. At the end of the millennium, there was an increased

© The Author(s) 2026

A. Dinar et al., *Water Institutions and their Performance*, SpringerBriefs in Water Science and Technology, https://doi.org/10.1007/978-3-032-24649-3_1

"

recognition that water scarcity reached unprecedented levels, affecting international relations, economic performance, and the sustainability of the water resources base (Bromley and Anderson 2018; Dinar and Dinar 2016; Saleth 2011). Water institutions have been recognized to make a significant difference in the performance of water resources in both domestic settings and internationally shared water resources.

We observe several approaches employed in studies aimed at realizing the value of institutions in the water sector. The most common approach is the 'before-and-after' approach, which has been applied in numerous individual case studies (at regional, state, or river basin levels) around the world. These studies compare different variables that measure efficiency or welfare before and after the presence of various water institutions (Araral and Ratra 2016). Another approach, also found in the literature, employs regression analyses to estimate the value of various water institutions (Bromley and Anderson 2018). Meta-analyses have also been used to assess the role of water institutions and their performance across various locations, physical, and economic settings (Özerol et al. 2018; Apio et al. 2025).

The purpose of this book is to extend the basis of the literature surveyed on water institutions from 1995 and introduce new studies that have been published in the past 30 years (until 2025) with the following question in mind: "Does progress in recent developments of institutional economics help sustain water resources under global changes?" We analyze water institutions by analytically distinguishing their three dimensions: legal, policy, and organizational dimensions, as suggested by Saleth and Dinar (2004).

The rest of this book is organized as follows. Next, we explain how we collected the data for the analysis and the empirical considerations arising from the construction of the dataset we developed and the methodology we employed, emphasizing the scope and scale of the analysis. Chapter 3 presents the analysis results, including a description of the dataset, the scale and context of the institutions addressed, and progress made in the legal, policy, and organizational dimensions of water institutions. Chapter 4 concludes the book and provides suggestions for future research directions on water institutions.

Chapter 2
Materials and Methods

Abstract This chapter presents the different sources used for the selection of relevant publications (published in the different outlets between 1995 and 2025) for the review, and the analytical methodology used in the review process. The chapter describes the analytical framework in the basis of our analysis, namely, analytically distinguishing the three dimensions or domains of water institutions, i.e., legal, policy, and organizational. We hypothesize that more effective water institutions lead to better performing water governance and vice versa. And we introduce, global changes (climate change, globalization of information and trade) as factors impacting the water sector through alterations in the level and pattern of water demand and supply, as well as the water institutions in place. Thus, healthier water institutions will be able to sustain global changes. We argue that the effects of these types of impacts on water institutions (and the water sector) observed across countries, times, and sectors can be better understood by evaluating them at various scales and sectors and other specifications. This allowed us to distribute the existing literature on water institutions into nine categories (coined "bins") and the results below reflect the various bins we use.

2.1 Search Strategy, Study Selection, and Data Extraction

The review is based on publications from 1995 to 2025, including journal articles, book chapters, books, and agency reports, identified in various databases (e.g., JSTOR, AGRICOLA, Social Science Research Network—SSRN). In addition, we searched published papers in 12 water-related and institutional economics-related journals. These journals include *Agricultural and Resource Economics Review, Agricultural Economics, American Journal of Agricultural Economics, Australian Journal of Agricultural and Resource Economics, Ecological Economics, Environment and Development Economics, Environmental and Resource Economics, Journal of Environmental Economics and Management, Journal of Environmental Management, Journal of Institutional Economics, Natural Resources Forum, Water*

A. Dinar et al., *Water Institutions and their Performance*, SpringerBriefs in
Water Science and Technology, https://doi.org/10.1007/978-3-032-24649-3_2

Economics and Policy journal, Water Resources & Economics, Water Resources Research, and Water-MDPI.

In selecting a publication for inclusion in this review, we identified those publications that specifically include the words *"water institutions"* or *"water governance"* in the title or display *"institutions"* or *"governance"* in the keyword list. Once a publication was identified, two co-authors read its abstract. If it met the inclusion rule, it was placed in a special category that describes its content and focus (as explained in Sect. 2.2). After deliberation and cross-examination among the co-authors, 36 of these publications were identified as irrelevant to our analysis and were excluded. A total of 275 publications were identified for inclusion. A more detailed description of the dataset and the distribution of publications across the identified issues is provided in Chap. 3.

2.2 Methodology

The methodology used in this book is based on several premises that enabled us to synthesize the studies we reviewed into a set of topics addressing the increasing complexity of the water sector worldwide. First, we approached water institutions by analytically distinguishing their three dimensions or domains, i.e., legal, policy, and organizational (Saleth and Dinar 2004). All three dimensions characterize the institutions we identify. Second, we treat (overall) water governance as the outcome of water institutions, i.e., more effective water institutions lead to better water governance and *vice versa.* Third, global changes impact the water sector through alterations in the level and pattern of water demand and supply.

Global changes encompass population growth, economic development, technological advancements (which also impact economic development), shifting consumption patterns (affected by economic development and technology), economic, financial, and political crises, as well as climate change-induced natural disasters. Fourth, global changes can also affect institutional water changes through their direct effects on transaction costs in the water sector and indirect effects on other sectors connected to the water sector. Thus, global changes can affect institutional changes both through their effects on water demand and supply and through their impact on transaction costs, which determine the supply and demand for institutional reforms or changes. Finally, the effects on water institutions (and the water sector) observed across countries can be better understood by evaluating them at various scales and sectors. Considering these premises would allow us to organize the existing literature on water institutions into nine categories (referred to as "bins").

Chapter 3
Results

Abstract The results chapter is divided into several secondary subsections including a description of the dataset, the 'bin' concept which allow us to highlights the evolution of the field of water institutions toward more integrative and applied approaches, addressing the complexities of water as a cross-sectoral resource. The bins cover several aspects of water institutions, such as the scale of the institutions, sectoral institutions, multi-sector institutions, institutions in developing vs. developed countries, institutions developed to deal with different types of water, institutions to manage international water, institutions to deal with climate change impact on the water sector, types of institutional intervention, and finally, undefined institutional premises that were not covered in the previous bins.

3.1 Description of the Dataset

For this research, we included 275 studies published between 1995 and 2025. These studies are divided into nine main categories/bins: the scale of the analysis (69), sectoral institutions (82), multi-sectoral institutions (56), developing vs. developed countries (16), type of water (15), international water (49), institutions to deal with climate change (36), means of institutional intervention including legal, policy, and organizational reforms (109), and finally miscellaneous, including those undefined institutional premises (9).[1] Table 3.1 presents the total number of studies that were included in each bin. We should indicate that individual studies could address premises described by several bins. Therefore, the number of studies included in the analysis in the nine bins exceeds 275.

The cumulative number of studies over time highlights the trajectory of scholarly attention to water institutions. Figure 3.1 presents the cumulative number of studies

---

[1] The total number of studies reviewed by bins is 441, which exceeds the 275 studies reported as total reviewed, as explained in the next paragraph. We have discussed in details each multi-bins study only in the first bin to which it was associated with. Then, if that study was identified also in association with other bins, we addressed its relevance to the other bins, but in less details.

© The Author(s) 2026

A. Dinar et al., *Water Institutions and their Performance*, SpringerBriefs in Water Science and Technology, https://doi.org/10.1007/978-3-032-24649-3_3

Table 3.1 Distribution of the papers with institutional premises across the nine bins in our study

Bin description	Number of studies
1. The scale of the analysis (basin, country/cross-country, local jurisdiction)	69
2. Sectoral institutions (irrigation, residential, industrial, environmental)	82
3. Multi sector institutions (e.g. water-energy-food)	56
4. Developing vs. developed countries	16
5. Type of water (surface water, groundwater, wastewater, conjunctive use)	15
6. International water	49
7. Institutions to deal with climate change	36
8. Means of institutional intervention: legal (water rights, water quotas, standards), policy (pricing, subsidies [direct and indirect], taxes, trade), and organizational (user organization, decentralization, enhancing capacity)	109
9. Miscellaneous (undefined institutional premises)	9
Total papers included in the analysis	275

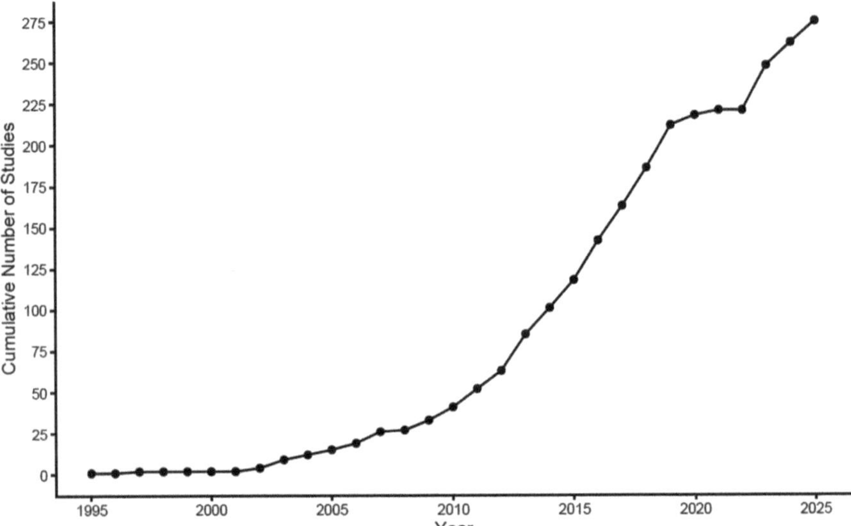

Fig. 3.1 Cumulative number of studies over time (1995–2025). (Source: Elaborated by the authors)

in the sample for the years during the period 1995–2025. In the early years (1995–2010) of the sample, publications were sparse, and the curve rose only slowly, reflecting the limited body of work and fragmented contributions. As the years progress, the curve steepens, showing a surge in research activity as the topic gains traction in academic and policy debates. This acceleration suggests both a growing recognition of the importance of water governance and institutional analysis and the maturation of methodologies and data sources that have enabled systematic study. In recent years, the cumulative curve has risen sharply, underscoring the

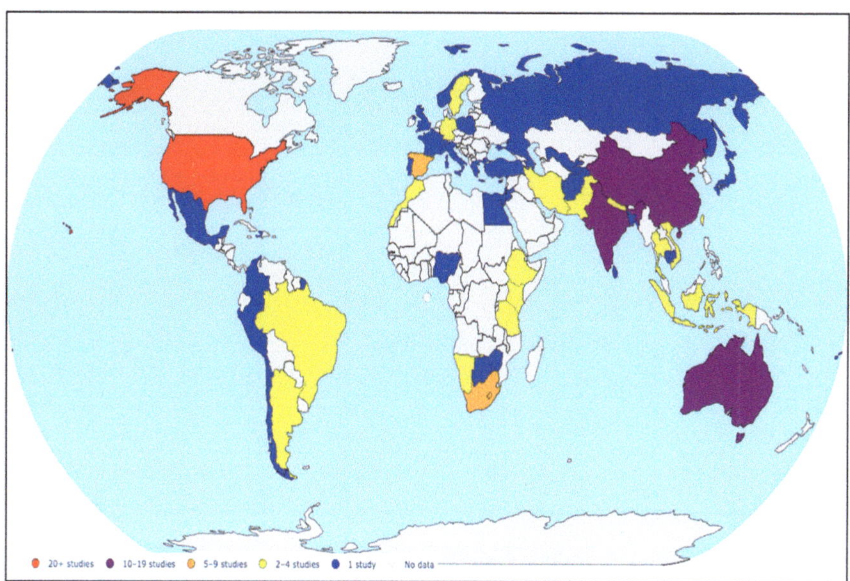

Fig. 3.2 Total number of studies by country. (Source: The map in this book was created using ArcGIS® software by Esri. ArcGIS® and ArcMap™ are the intellectual property of Esri and are used herein under license. Copyright © Esri. All rights reserved. For more information about Esri® software, please visit www.esri.com)

momentum of scholarly output and indicating that the field has transitioned from a niche area into a well-established domain of inquiry with significant breadth and depth.

The distribution of studies by country (Fig. 3.2) shows a clear geographical concentration of research efforts. The United States, China, Australia, and India stand out for the highest number of studies, reflecting both their substantial research capacity and the pressing water governance challenges they face. Countries such as South Africa, Spain, Germany, and Kenya also contribute a notable share. In contrast, many countries in Latin America, the Middle East, and parts of Africa are underrepresented, or not represented at all, suggesting either underrepresentation in the academic literature or challenges in data availability. This uneven distribution highlights important gaps in global coverage, where lessons from well-studied contexts may not fully capture the institutional complexities of less-studied regions, underscoring the need for more geographically diverse research.

The analysis of studies by binning them over time (Fig. 3.3) reveals how scholarly attention has been distributed across thematic areas of water governance. Certain bins, such as sectoral institutions and scale of analysis, consistently account for a substantial share of studies, reflecting the foundational role of institutional arrangements and governance scales in shaping water outcomes. Other bins, such as international water and institutions dealing with climate change, have expanded their shares in the annual studies in recent years. The bins related to multi-sector

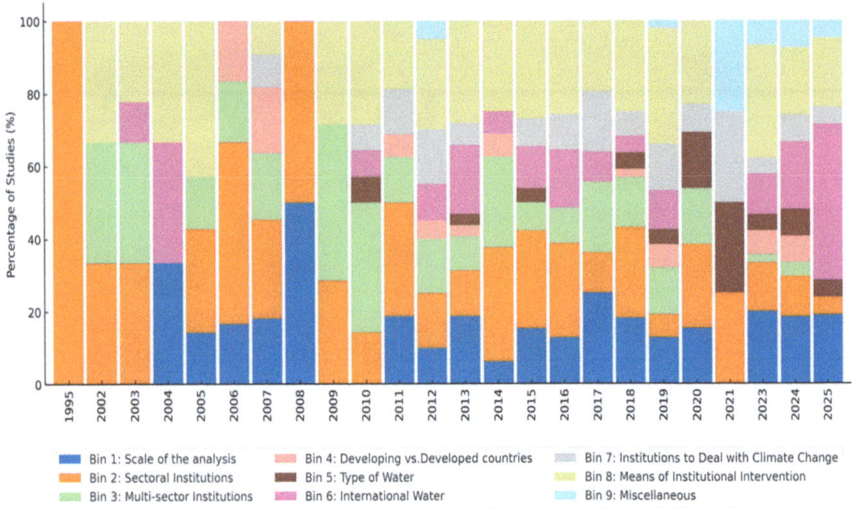

Fig. 3.3 Share of studies by bin over time. (Source: Elaborated by the authors)

institutions and means of institutional intervention highlight the evolution of the field toward more integrative and applied approaches, addressing the complexities of water as a cross-sectoral resource. Overall, the bin-based distribution illustrates how the field has expanded from a narrow focus on traditional governance mechanisms toward a more diverse, multidimensional research agenda, reflecting the complexity of water challenges in different socio-ecological contexts. In Sects. 3.2–3.10, we summarize papers in each of these nine bins.

3.2 Scale of the Analysis

Understanding the scale of analysis is essential in institutional studies of water governance, as the effectiveness of frameworks, policies, and institutions often depends on whether they are applied at global, national, regional, or local levels. Not all governance mechanisms function equally well across different scales; some are designed for broad coordination across countries, while others are more effective through close engagement with local communities and users. In reviewing 69 papers, we categorize them by their primary scale of analysis: global, cross-country comparative, national, subnational or river-basin, and local or community. This organization helps clarify not only the spatial focus of each study but also the types of institutional challenges and innovations that emerge at different governance levels.

3.2.1 Global Scale

At the global scale, Gupta and Pahl-Wostl (2013) explore the challenges of global water governance and address whether water challenges are global challenges and what type of governance and water resource management would be appropriate. The article suggests that management at the river basin level is now being overtaken by global water governance. Similarly, Hoekstra et al. (2019) point out the use of water footprint assessment in promoting collective action in water governance, reviewing ten different studies and their use of the water footprint assessment (WFA). The papers reviewed address a wide range of topics, concentrating on different scales of governance and different phases of the product supply chain. Their findings also show that understanding actions and players at different spatial levels along supply chains is essential for the WFA to perform more sustainably.

Furthermore, Makate et al. (2018) review the water footprint methodology to increase stakeholders' understanding of the concept, its implementation, and its relevance in water use and management. The article argues that the water footprint approach is an effective policy instrument for managing limited freshwater resources and fostering sustainability and governance in human-induced water usage worldwide. More importantly, the idea can help disengage development from excessive water resource usage and its environmental consequences. As a result, it is critical for many stakeholders in the developing world to understand and use the water footprint idea to minimize "excessive physical and economic water scarcity." In a related critique, Hoekstra (2011) expresses how the River Basin Approach is not always sufficient. Interestingly, this article highlights four vital worldwide concerns that need to be addressed: efficiency, equity, sustainability, and security of water supply in a globalized world. Finally, Gopalakrishnan et al. (2005) offer a broad global survey of institutional arrangements across various regional and international contexts. It includes an analysis and discussion of the rationale for institutional innovations in Hawaii, Sri Lanka, India, China, Mexico, and several international comparisons in the Middle East and the Western United States.

Wutich et al. (2012) refer to tensions between institutions and existing cultural norms of justice. The paper proposes a new approach to cross-cultural analysis that can investigate these tensions, which can then assess when local cultural norms are likely to facilitate or impede the acceptance of specific institutions. The approach is implemented, using data collected from respondents from five global sites (in Fiji, Ecuador, Paraguay, New Zealand, and the U.S.) using cultural consensus analysis. Findings suggest evidence of similar cultural norms of justice in water in the following domains: human right to water, water governance, water access, environmental stewardship, aspects of water markets, and aspects of water quality and health.

Ibrahim and Farah (2025) examine the international frameworks that were developed separately for climate change and for water security concerns, for both shared surface and groundwater sources. The paper promotes an improved interaction between international climate change law and international water law to recognize the synergies between climate change and water governance at both global and basin levels. The proposals to enhance synergies between international climate change law and international water law are applied in the Nile River Basin and the Guarani Aquifer.

3.2.2 Cross-country or Comparative Analyses

Cross-country studies provide valuable comparative insights into how water institutions function under diverse political, economic, and legal systems. These studies often identify structural patterns, regulatory challenges, and institutional successes or failures that emerge when comparing different governance models. By evaluating water management strategies across continents, they highlight transferable lessons and the need to tailor institutions to socio-economic and cultural contexts.

Camkin and Neto (2016) use a set of essential water issues to analyze the rights and duties of diverse players in water governance. This research can be considered at either a cross-country or local jurisdictional level due to its scope. Building on this comparative perspective, Muchapondwa et al. (2018) aim to identify successful market-based incentives to help promote sustainable watershed practices through strengthening and increasing direct participation by local communities and the private sector. To do that, the paper reviews a sample of 26 case studies from Africa, Asia, Europe, and the Americas. Among the attributes compared are the threats to specific watersheds, the market-based incentives used, the countrywide policy environment, the outcomes from the interventions, factors explaining success and failure, and the pertinent policy issues in support of upscaling and expanding the appropriate market-based approaches.

In contrast, Golovina et al. (2023) investigate the organizational, economic, and regulatory aspects associated with groundwater extraction by individuals in the Russian Federation and compare it to the situation in Germany and China. The paper reviews the state-level legislation and regulation related to individual groundwater pumping and identifies shortcomings in the system, suggesting alternative mechanisms for legalizing the activities of individual groundwater users.

Sosa and Zwarteveen (2014) review literature on water and mining in Peru to assess the effectiveness of institutional instruments for safeguarding the sustainability of water resources and the water-dependent ecosystems in mining regions. The paper illustrates the approach and findings, using the Yanacocha gold mine in Peru. Water sustainability is regulated through "upward forms of accountability", where the mining company has to comply with national and international regulations, but local communities are not considered. The paper calls for the introduction of enabling institutions to consider the experiences, rights, and knowledge of communities affected by mining.

Ménard (2017) reviews the variety of arrangements, coined as 'meso-institutions', which are used to regulate water utilities, and identifies the mechanisms through which the general macro-institutional interact with operators organizing transactions at the utility level. The analysis draws on recent developments in organization theory and the distinction between property rights and decision rights, and applies it to drinking water utilities in France, England and Wales, and the Netherlands.

Taylor and Eberhard (2020) present a comprehensive review of the successes and failures of various institutional arrangements to protect the Great Barrier Reef (GBR) in Australia from water pollution damages by agriculture. The paper identifies the influence of factors such as diverse landholder goals and objectives, social and cultural objections to participation in protection programs, and the value of knowledge exchange over knowledge transfer in building trust and commitment amongst landholders. The review also highlights the importance of maintaining collaboration and the capacity of all stakeholders.

3.2.3   *National Scale*

National-level analyses focus on how institutions function within a country's formal governance system. These papers explore national reforms, governance failures, implementation gaps, and policy instruments. They often address issues such as property rights, subsidy distortions, and political constraints that shape institutional effectiveness.

Ananda and Aheeyar (2020) apply a water rights framework analysis to compare the performance of all groundwater aquifers in India. Their analytical framework (a multi-dimensional property rights model) allows them to identify faults and options for improvements. For example, there is much scope for promoting the 'small group groundwater sharing' arrangements. The work identified that distortionary subsidies (agriculture, groundwater) have negative impacts on the sector and the resource.

Similarly, Berardo et al. (2013) evaluate the development of several institutions by using the adaptive governance framework applied at the national level. The analysis allows for assessing the ability of national and provincial authorities to face the challenges of representation and decision-making that are critical in institutional development in the water sector. Bruns (2017) in turn, studies polycentric water governance in Southeast Asia and provides evidence that polycentric water governance may take several forms, including adaptive co-management in irrigation, weaving network governance, and practical problem-solving in river basins, all of which can be developed through institutional improvisation. However, attention to governance failures is also critical to understanding the sustainability of water supply.

In contrast, Bakker et al. (2008) highlight institutional failure in Indonesia. This study focuses on governance failure in the institutional dimension of a project to supply urban water to poor households and found that while the project is aimed to provide new, subsidized water supply connections to 20,000 families; yet, most houses that qualify as sufficiently impoverished to get a subsidized connection do not have legal land tenure status, preventing them from participating in the program. Although private-sector operators sought to lobby the government to modify the tenure policy, the initial target of 20,000 has been reduced to fewer than 4000 families.

Shah and Narain (2019) take a more critical stance, arguing that national narratives about water scarcity in India often obscure the institutional causes of inequality and mismanagement. The authors call for a focus not only on resolving hydrological unavailability through supply-side solutions, but also on the underlying institutional causes that drive water inaccessibility.

Meinzen-Dick (2007) examines the experience of searching for panaceas in the irrigation sector. It reviews the evidence on policies promoting institutions such as government agencies, user organizations, and water markets. A study of the variable performances of user groups for canal irrigation in India illustrates the factors that affect institutional performance and reveals that each of these institutions has worked in some situations but failed in others.

Tortajada and Joshi (2013) analyze Singapore's national water management system, emphasizing the importance of integrating planning, development, and governance into a single institutional vision. The authors argue that including these components separately in any framework does not automatically guarantee improvements in water-sector performance unless there is a strong emphasis on integrating all components described above into policy development and program implementation.

Falk et al. (2019) use experimental games to help better understand coordination challenges and establish institutional capacities related to managing small village reservoirs in India. By combining game playing with group discussions of possible institutional solutions to cooperation patterns, the participants (300 water managers) performed considerably better than standard economic theory predicted and also substantially better than observed real-world behavior. The discussions showed that the players connected the game to their real-life experiences and understood the performance of the evaluated institutions.

Ávila et al. (2023) address the functionality of water supply and Sanitation services in Brazil, based on the Institutional Analysis and Development (IAD) Framework. Using national-level data, the paper evaluates existing structures for providing WSS to municipalities in Brazil's large metropolitan regions. Results clarify the actors in the WSS, their functions and responsibilities, and their interactions, as well as the impacts of various exogenous factors. An interesting aspect of this paper is the use of a Delphi approach in deriving recommendations.

Söderberg (2016) analyses Swedish water governance as a case to enhance the understanding of policy implementation in complex governance structures. The paper maps out the formal structure of the water governance system of Sweden, focusing on power directions within the system and analyzing policy coherence.

3.2.4 Subnational/Regional/River Basin Scale

This group of studies explores governance mechanisms and institutional arrangements at the level of river basins, provinces, or watershed regions. They examine adaptive governance, inter-jurisdictional coordination, water rights allocation, pricing policies, and ecological responses to institutional reforms. These works often illustrate the complexity of scaling policies and the challenges of achieving integration across administrative and hydrological boundaries, particularly in trans-boundary or shared basins.

Garrick et al. (2011) examine the institutional innovations to govern environmental water in the Western United States. The findings conclude that in the Western United States, two decades of implementation experience offer a middle ground between top-down and bottom-up approaches: Layered governance frameworks for basin-scale integration and accountability that invest in local institutional capacity while guaranteeing to complement state and federal duties. Building on this regional focus, Grigg (2018) also examined collective action in multilevel water governance and management in the Western U.S. and demonstrated that the magnitude or scale, as well as the nature, of the water management challenge influence collective action. Collective findings from the cases and examples in the study demonstrate that as scale increases, empowerment decreases in diverse ways, with greater empowerment for planning issues, less for infrastructure development issues, and even less for regulatory actions. Problem scales can exceed the micro watershed level, which often means that individuals will need to express their needs through the voice of other representatives. This article stresses that management and governance must consider the different scales individually, which will help resolve water conflicts and problems.

Meanwhile, Layzer and Schulman (2013) evaluate the adoption of the IWRM in the U.S., using the Chesapeake Bay Program (CBP) as an example, which is considered the nation's premier watershed-scale management initiative. Based on journalistic sources, in-depth interviews, and review of program documents, the authors conclude that the CBP has enhanced technical, institutional, and—to a lesser extent—sectoral integration in the entire watershed. However, due to the cooperative and voluntary nature of the program, as well as the lack of adequate resources, the program's goals, primarily reforming existing behaviors, have not been fully met, and ecological improvement in the Chesapeake Bay has been minimal.

Roßner and Zikos (2018) study focuses on water governance in Uzbekistan, using an economic-framed field experiment to assess the impacts of externally enforced and self-set water distribution regulations on homogeneous and heterogeneous groups of irrigation users with varying economic endowments. Twenty farmers from an Uzbek village participated in the study and were interviewed, complementing the analytical method. The findings suggest that economically homogenous organizations are more likely to follow self-designed norms than economically diverse groupings. As a result, under self-governance, homogeneous groupings do better in terms of resource conservation and water harvesting.

Similarly, Cody (2019) applies multi-level selection (MLS) theory and CPR theory to a stratified, semirandom sample of 71 irrigation systems of distinct cultural origins in the Upper Rio Grande Basin in the US to test hypotheses related to the role of norms in irrigation system performance. Findings suggest that systems with internalized norms of cooperation have governing rules and more advanced technologies, leading to better care for the commons, public goods, and higher equality among irrigators.

Shivakoti and Bastakoti (2006) compared the dynamics and robustness of two irrigation projects in the Kok River in Northern Thailand within the Mekong River basin in the context of changing governance mechanisms and evolution of technological and market shocks. The user group efforts, their self-governing capabilities, and local institutions, including the changes that took place in the role of the leaders in response to the external shocks, have played an important role in the robustness of the performance of these projects. The paper concludes that policymakers should give due attention to existing local institutions, including their long-term ability to respond.

In East Africa, Kuhn et al. (2016) use the Lake Naivasha Basin in Kenya's Rift Valley as a hydro-economic system with slowly emerging basin-wide water management institutions to assess the institutional needs of this system. The paper refers to the two main issues the basin faces, namely the large-scale horticultural activities requiring substantial and regular amounts of irrigation water, and the volatile weather conditions where periods of average and above average rainfall have alternated with prolonged drought periods. The paper analyzes the effect of existing and proposed water institutions on preserving target lake levels during periods of highly volatile water availability.

Extending to Southern Africa, Söderbaum (2015) offers an alternative way to think about managing and governing an international river basin. They address the challenge to reconcile national benefits and interests with the common good and a basin-wide approach. Using the case of the Zambezi River basin, they raise the question of whether transboundary waters are best governed through specialized and functional river basin organizations (RBOs) or through more multipurpose regional organizations that have greater distinct political leverage.

Satoh (2019) examines the effectiveness of the water rights institution in Japan's Abukuma River region. Findings suggest that inefficiency arises when water suppliers with water rights face declining demand, but only when excess water rights are reallocated to increase the efficiency of the water suppliers. This suggests that the government should permit the reallocation of water rights under a trading scheme, leveraging market institutions to address differences in efficiency between water rights holders.

Gerlak (2004) evaluates activities undertaken by the Global Environmental Facility (GEF) in the Danube River Basin. While the GEF was successful in developing scientific knowledge and strengthening regional governance bodies in the various riparian states in the basin, coordinating across the growing number of participating groups in such a large basin and their recognition of the positive value of obtaining such knowledge presented GEF was successful in developing scientific knowledge and strengthening regional governance bodies in the various riparian states in the basin, the coordination across the growing number of participating groups in such a large basin and their recognition of the positive value of obtaining such knowledge was a challenge. The paper identified several issues that are critical for future river basin institution building and decision making: (a) the importance of appropriately creating and disseminating scientific data pertaining to the river system, (b) the need for basin-wide governance bodies for integrated river basin management, and (c) the necessity to address coordination issues throughout project planning and implementation.

In the Australian context, Wyborn et al. (2023) introduce the concept "polycentric spectrum" to demonstrate the difference between ecosystems that resist change and those that are easier to modify upon institutional reforms. The paper applies the concept to the Murray-Darling Basin (MDB) in Australia. The paper argues that for a reform to be more easily achieved, the reform agenda has to be detached from the perception that the government is the sole agent of change. Still, there needs to be strategic efforts to engage local- and non-state-level actors who play a central role in the sector and can help adapt the necessary water governance changes. Lai and Zhao (2023) apply an institutional analysis and development framework to identify the actual operating rules of Australia's MDB. The results provide insights regarding the key actors in the MDB's institutional structures and their characteristics. These include the position, boundary, choice, aggregation, information, scope, and payoff rules that regulate the water management in the MDB.

Syed and Choudhury (2018) apply the Transboundary Water Interaction Nexus (TWINS) approach to the Indus Basin in Pakistan, focusing on multi-scalar governance of a transboundary river. The paper argues that studying cooperation and conflict at the international water level, addressing multiple scales and multiple stakeholders, is required to reach the degree of cooperation and to ensure the effectiveness of transboundary governance structures. Using the TWINS approach applied to the Pakistani part of the Indus, the inclusion of scale allows for the accounting of the complexity of transboundary water governance and the design of effective governing institutions.

Oberlack and Eisenack (2018) conduct a meta-analysis of 26 global case studies to identify barriers to adapting river basin governance because of climate change. The authors introduce an archetype framework, categorizing identified barriers into patterns, such as coordination gaps, path dependencies, competing priorities, and more. The results emphasize how adaptation barriers are also caused by the collective action of pre-existing institutional arrangements, challenging the conventional focus on biophysical features.

Kajisa and Dong (2017) evaluate the impact of volumetric water pricing on the performance of different management institutions of farmers' groups applied to water user associations in the Hubei basin in China. The work provides evidence that proper volumetric price design sends a signal for institutional change for reservoir water savings.

Villamayor-Tomas (2018) examines how irrigation associations respond to climate shocks, showing their institutional flexibility at the basin scale. The paper demonstrates how irrigation associations and other local natural resource management organizations in different countries are institutions well-suited to address external shocks affecting water users. Janssen and Anderies (2013) apply a multi-method approach, including case-study analysis, experimental methods in laboratory and field settings, and mathematical models. They synthesize lessons learned from a series of studies on small-scale irrigation systems to show the importance of creating institutional arrangements that fit the human ecology within the biophysical constraints of the system.

Hayden and Tsvetanov (2019) evaluate the effectiveness of the outdoor irrigation water restriction institution using information from 408 urban water suppliers in California during the 2012–2017drought. The findings indicate a minor change in water consumption resulting from the restriction of one day a week, but with large heterogeneity in the impact. They find that the effectiveness of this institution declines as restrictions tighten. Additionally, this institution performs more effectively in regions where residential water use accounts for a larger share of total urban water consumption.

Gerlak (2017) emphasizes the importance of participation and engagement of stakeholders in river basin institutions. While good and functioning institutions are necessary conditions, the need for participation is no less important. The paper outlines the growing importance of public participation and stakeholder engagement in river basin institutions in the Colorado River Delta, to uncover an emerging network governance.

Lawless et al. (2024) examine the changes in water management strategies in the Colorado River Basin between 1922 and 2022, using institutional, temporal, and network structure analysis methods. The water regulatory network in the basin was evaluated at several points in time to identify changes to the institutional structure, actors, and governance level. Findings suggest that changes took place to constitutional, operational, and collective-choice level rules for water supply, storage, conveyance, and use. The changes were made by adding layers of new governance rules without introducing major changes to responsibilities assigned to decision-making positions, which have remained, and actors who issue and are targeted by the rules have undergone significant changes.

Handayani et al. (2023) examine the water governance framework for the city of Semarang, Indonesia, using documented stakeholders' engagement in water management, and interactions among themselves and with levels of government. Stakeholder roles analysis and Social Network Analysis (SNA) were applied. Focus Group Discussions and interviews were used to document the dataset for the analysis. It was found that while the city of Semarang has succeeded in creating an enabling involvement of various stakeholders, it is still being challenged by its capacity to operate adaptive water governance, resource mobilization, and reducing dependency on the National Government and International financial Agencies.

Sithirith et al. (2023) evaluate the institutional framework that manages the 39 river basin systems in Cambodia. Moving between excessive water in the wet season and insufficient water in the dry season, floods and droughts have a negative impact on the country's water economy. The overdevelopment of infrastructure, irrigation schemes, and hydropower facilities in the Mekong region by other riparian states (China, Thailand, Vietnam, and Laos) altered the hydrological regimes of the Mekong River and its stem in Cambodia. The paper calls for better coordination of national interests, power relations, and upstream and downstream politics, or in other words, regional institutions that are based on regional cooperation frameworks.

Suleymanov (2024) reviews the institutionalization of the Kura-Aras River Basin, an international water-scarce basin in the South Caucasus with Azerbaijan, Georgia, Armenia, Iran, and Turkey as its riparian states. The paper used a semi-structured interview framework to collect data from experts in the various riparian states regarding basin-wide institutions to allow effective water resource management. Results suggest that interviewees across the basin states agree on the importance of establishing robust governance mechanisms that foster cooperation, enhance transparency, and mitigate conflicts among riparian states. However, the lack of trust between riparian states has been identified as a potential for regional non-cooperation.

Sehring et al. (2024) explore the politics of institutionalizing river basin management in Central Asia, the Former Soviet Republics, with a focus on the establishment of basin management organizations in Kazakhstan and Tajikistan. The authors use published academic literature, policy reports, and interviews with experts in the region. Results show a close interaction between the interests and activities of national and international actors, and that the influence of water governance norms promoted by international donors has initiated legal changes in all countries of the region, though implementation has been uneven. The results of this paper overlay those of Suleymanov (2024), which refers to a different basin in the same region, and finds similar reasons for failing to promote functioning institutions.

Prniyazova et al. (2025) examine the complexities of water resource governance in the Central Asia/Aral Sea region, emphasizing the relationship between national interests and regional cooperation. The paper demonstrates how water scarcity, climate change impacts, and the growing tensions over transboundary river basins in the region affect the linkages between water, energy, and agriculture among riparian nations. Existing water-sharing and management agreements remain fragmented despite diplomatic efforts to shift the riparian states toward enhanced regional cooperation in water and energy.

Kibaroglu (2025) reviews formal and informal water institutions in the Euphrates-Tigris Region, including Turkey, Syria, and Iraq. The transboundary water relations in the basin are marked by political confrontations among all riparian states, yet the basin also hosts a complex range of formal and informal water management institutions that operate at varying levels of effectiveness. The chapter critically analyzes water institutions in the Basin with a focus on the current institutional frameworks, such as bilateral high-level political and bureaucratic dialogue between Turkey and Iraq, as well as the science-diplomacy and capacity development initiatives that have been developed during the prolonged crisis in the region. The chapter also analyzes the evolution of national water management institutions related to river basin planning and sectoral (i.e., irrigation) water policy and management issues.

Bitterman et al. (2023) apply a network analysis of water governance in the Lake Champlain Basin in Vermont. Using a survey of water governance actors in the Lake Champlain Basin, the paper maps the structure of the water governance network and identifies the key information brokers, flows of resources, and ongoing collaborative partnerships. This approach allows measurement of cross- and within-scale linkages that characterize the degree of coordination across space and scale. Findings suggest that as the governance system transformation proceeds, cross-basin and inter-basin coordination needs to be regularized to sustain learning and innovation across the basin system.

Colby and Reed-Spitzer (2024) evaluate, using economic criteria, features of eight water agreements that have occurred in the Southwestern United States between national and state institutions for addressing conflicts over water entitlements of Native American tribes. The evaluation suggests that recent agreements perform better on the economic criteria, implying that learning about effective agreements is occurring over time.

Gharib et al. (2024) developed a methodology that evaluates how population location under alternative water institutions and climate scenarios impacts water demands, shortages, and derived economic values. The methodology is then applied to the South Platte River Basin in Northeastern Colorado. It was found that while water rights institutions have a negligible impact on total volumetric shortages relative to climate change, they have substantial distributional and economic implications.

Marshall et al. (2017) highlight the interaction between institutions and human and technical capacities in managing international water. Their main argument is that institutional and human capacities often need to be restructured to enable effective implementation and monitoring across transboundary or international water systems, reinforcing the importance of capacity-building at regional governance scales.

Zikos and Roggero (2013) perform an analysis of interview data about Cyprus's water conflicts, arising from the geopolitical split between the island's two communities. The study contrasts two institutional logics: "island fit," which supports unified, island-wide governance; and "patronage fit," which aligns each community with external patrons (Turkey, Greece). The analysis reveals a preference among Cypriots for unified institutional arrangements, reflecting a shared sense of identity. This study fits squarely within the regional scale as it addresses the governance of a divided territory and the competing frameworks for basin-wide water governance across a politically fractured island.

Martín Velasco et al. (2023) apply the OECD Water Governance Indicator Framework at the General Pueyrredon Municipality in Buenos Aires Province, Argentina. They assess institutional gaps at the subnational level, particularly in translating legal frameworks into actionable water management practices. Despite strong coordination between legal frameworks and enforcement institutions, challenges persist in the operational "how" of water governance. This analysis exemplifies subnational governance concerns and highlights the need for implementation-focused capacity building within existing institutional frameworks.

3.2.5 *Local/Community/Utility Scale*

Local-scale studies highlight how institutions operate at the grassroots level, in small communities, irrigation districts, user associations, and utilities. These studies focus on stakeholder engagement, cultural norms, capacity constraints, training effectiveness, and the dynamics of self-governance. They show that while formal rules matter, informal norms, trust, leadership, and local knowledge often determine the success or failure of institutional arrangements at this level. They also emphasize the need for locally appropriate, flexible, and participatory governance.

Stoa (2017) explores water governance in northern Haiti and finds that, while local stakeholders are engaged, human and financial resources are inadequate to fulfill statutory duties, according to a capacity evaluation of institutions in northern Haiti. The findings imply that local governments and community institutions should be included in water resource management planning, with national or international assistance augmenting capacity and technical support.

Tiger et al. (2011) analyze the role of measuring water use for indoor and outdoor purposes as a means to improve monitoring and regulatory institutions of household conservation. The paper argues that utilities should promote the installation of irrigation meters by offering financing options for upfront installation costs. When setting initial and ongoing costs associated with dedicated irrigation meters, utilities must weigh the value of the metering and monitoring.

Balasubramanya et al. (2018) assess the impact of training on water user association (WUA) staff on the performance of mandated duties of the WUA. Using 74 WUAs in Tajikistan that were created using a longer training process, with 67 'control' WUAs that were created using shorter training. The methodology employed generates evidence that, while biased toward generating an underestimate of effect, can still be useful and informative for policy and management purposes, and for evaluating the impact of the training process on the functioning of new institutions in transition settings.

Behera and Mishra (2018) evaluate the performance of the long-known water management institution, Panchayat. The authors find that the Pani Panchayats evaluated, which have been supported by the Odisha (Orissa) Pani Panchayat Act (2002) and follow a democratic decision-making process, have been effective in managing the different irrigation projects efficiently, addressing water pricing and distribution of irrigation water to users, maintenance of the systems, and their expansion. All of that has to be socially inclusive and sustainable agricultural development in the analyzed regions.

Franzén et al. (2015) compare institutional arrangements in two Swedish catchments, showing how traditional institutions affect organizational arrangements and stakeholder participation in the watershed. Using a literature review and semi-structured interviews, the results indicate a preference for different institutional arrangements for water management, despite similar physical conditions and the same national- and regional-level regulatory framework.

Schnegg and Bollig (2016) examine the ability of water user associations in rural communities in Namibia during drought years to administer water services. Using a long-term ethnographic fieldwork in seven communities, they examined whether and how water management regimes were either modified or applied. The findings suggest that cultural models of kinship and reciprocity took priority over formal agreements during the drought, showing non-adherence to formalized institutions.

Nyagumbo and Rurinda (2012) evaluate the effectiveness of various water management institutions for smallholder farmers in Zimbabwe. The data they use are based on literature reviews, stakeholder workshop feedback, interviews with key informants, and evaluations of policy impacts across various case study projects/programs. The findings from the analysis suggest that the prevailing institutional structures that were reformed since the 2000s were not effective in delivering services to smallholder farmers due to a lack of human and financial capital on both sides of the agencies and the farmers themselves.

Babbitt et al. (2015) apply mixed qualitative and quantitative data collection efforts to look at how a localized and integrated approach to managing the water resources management system operates. Analyzing data provided by stakeholders living and operating in the basin suggests that overall, the current water management system is performing relatively well, even though it is still in its infancy. The analysis also found that performance could be further improved by ensuring all stakeholder interests are represented.

Mwakalila and Muneeni (2025) investigate the effectiveness of local water institutions' management of water resources in the Sanya-Kware sub-catchment of the Pangani River Basin, Tanzania. The paper examines the advantages and limitations of traditional 'furrow committees' and their interactions with newly established Water User Associations. Results reveal that institutions based on the 'furrow committees', influenced by cultural norms, provide effective, localized solutions for water allocation and conflict resolution, but face challenges beyond field-level management, such as limited Application of Integrated water resources management principles, a lack of legal enforcement, and insufficient capacity for investment decisions.

Sowby and South (2023) compare two utilities whose novel pricing structures reflect water availability and reasonable use expectations. One has a physical basis in water rights, landscape water needs, and irrigated area; the other has a hydrologic basis in an annual water-supply forecast. Both utilities charge users tiered (increasing block) rates for an adjustable water allotment. The rates have proved hydrologically beneficial, socially equitable, and financially sustainable during the drought, suggesting that water pricing can be an effective policy tool for responsive water management.

Olivier and Vallury (2024) combine aspects from policy design and from common-pool resource governance to assess the extent to which Nebraska's Natural Resources Districts (NRDs) design plans and programs that fit their social-ecological contexts. NRDs were created to provide context-specific solutions to local water problems. Using text analysis and Qualitative Comparative Analysis of published plans and programs created by NRDs suggests that the biophysical context plays a role in shaping plan and program content, but that broader, top-down institutional mandates may be more significant in shaping the outputs produced by NRDs.

Araral (2009) models the incentive problems in foreign aid, focusing on how moral hazard and aid fungibility interact with incentives faced by public agencies and the recipients of the aid in developing countries. The model is applied to the irrigation systems in the Philippines, which are regulated by the National Irrigation Administration (NIA). The findings are broadly consistent with the theoretical expectations of institutional rational choice. That is, the NIA has strong incentives to underspend in maintenance because it would re-direct constant funding that otherwise could be used for the agency's internal needs.

Bettini et al. (2015) address the phenomenon of transitional management in urban water utilities (e.g., build and transfer). The paper first develops the necessary analytical institutional framework to explain transitional management, and then presents a method for its empirical application. The authors argue that the tool they developed, using systems analysis methods, is likely to improve the transition management of water utilities by expanding empirical research.

Brent (2017) evaluates the economic value of secure water rights at the district level in Washington's Yakima Basin, finding heterogeneity in how secure rights influence property values. The relative value of secure property rights is affected by the water supply volatility, and the costs of drought damages are predominantly borne by those with less secure (junior) rights. The analysis did not find that secure water rights end up in higher property values at the irrigation district level, while there is still heterogeneity in the premium for secure water rights.

Berge and Torsteinsen (2023) explore how too much governance may influence governance control of municipal service-provision, where the principal and the agent operate under different institutional logics. The application of a principal-agent model to the urban water services in Norway suggests that the relationship between formal institutional structure and institutional logics is reciprocal, leading to antagonism between the principal and the agent, and thus malfunctioning of the organizational system that was created, instead of supporting it.

3.3   Sectoral Institutions

Eighty-two papers describe the sectoral institutions' section. Water institutions do not all belong to the same category, and different water institutions are divided into sectors depending on the type of water use. Most of the papers we reviewed belong to five main sectors: agricultural irrigation, residential (urban/municipal), industrial, and environmental. Common themes repeated across the papers include institutional performance in environmental water allocation, irrigation and agriculture, urban water, wastewater, water governance, water management, integrated regional water management (IRWM), incentives, and water rights.

3.3.1 Institutional Performance in Environmental Water Allocation

A prominent theme across the literature is the institutional performance in environmental water allocation. Garrick and Aylward (2012) examine how transaction costs impact the institutional performance of market-based environmental water allocation in the Columbia Basin, which historically has over-allocated for agricultural use. They specifically focus on the costs of defining, negotiating, and enforcing water rights transfers, which create greater barriers to effective environmental water recovery. By highlighting institutional capacity through an analysis of water recovery rates, Garrick and Aylward introduce the concept of "adaptive efficiency", emphasizing the need for institutions that can manage complex situations and improvise well, especially in ecological water management. Similarly, Garrick et al. (2011) analyze the evolution of institutional frameworks governing environmental water allocation in the Western United States.

Mogomotsi et al. (2018) explore Botswana's water supply institutions and allocations, showing how their water scarcity results from physical barriers and outdated institutional frameworks for sustainable management. By applying an institutional economics lens to analyze formal water governance structures, the authors describe how sectoral regulations and general policies contribute to inefficiencies in water allocation, balancing both sectoral and environmental demands. Despite Botswana's success in improving water institutions for the rural and urban sectors, they have delayed the implementation of reforms like the Integrated Water Resources Management (IWRM), leaving its water governance unable to handle cross-sectoral challenges. They emphasize how participatory governance, clarity of rights, and adaptive frameworks can support sustainable water management that integrates environmental and other sectoral needs.

Binz et al. (2016) investigate the institutional processes of the legitimization of reusable potable water in California, which posed many social and regulatory challenges at the time. They develop an analytical framework that combines the technological innovation system theory and institutional work concepts, which helps explain how different industries, regulatory bodies, and utilities collaborated to legitimize reusable potable water through strategies like advocacy and public education. Through the process of legitimation, they also supported sectoral water management integration and sustainable water supply diversification, addressing environmental concerns, residential needs, and agricultural needs simultaneously.

Gaden (2016) examines transboundary governance of the Great Lakes fishery, which is a unique sector managed by multiple American states, Native American tribes, and Canadian provinces. They developed the Joint Strategic Plan for Management of Great Lakes Fisheries to ensure cooperation and coordination through lake committees that use empirical and policy-based information to make decisions. In contrast to urban and irrigation systems' local scales, this governance model balances sub-national autonomy with collective action. The authors describe the usage of four institutional indicators, which are functional intensity, stability, legitimacy, and compliance, illustrating how this management strongly collaborates despite there being multiple large authorities.

Plumb et al. (2018) explore how irrigation districts in Oregon, which operate as local common-pool resource management (CPRM) institutions, respond differently to Payment for Ecosystem Services (PES) programs that incentivize voluntary water transactions for instream flows. By conducting qualitative interviews with district managers, the authors create a categorization of three district responses, which are water rights protectors, cautious converters, and new pioneers, based on variables like adaptability and infrastructure. They emphasize that institutional arrangements, such as investing in infrastructure, considerably shape how districts perceive, facilitate, or prohibit water leasing for environmental goals.

Peat et al. (2017) examine how water management agencies can create institutional flexibility to effectively implement adaptive management for aquatic ecosystem restoration, comparing the practitioners in Australia's Murray-Darling Basin to those in Southern Florida. The authors find that both agencies created limited institutional flexibility through conservative temporary agreements and collaborative leadership approaches, but informal institutional arrangements often weakened adaptive management more than formal rules and policies. While agencies successfully implemented small-scale active adaptive management experiments to build trust and test restoration approaches, the results indicate that most management occurred through passive adaptive management at larger scales.

Wan et al. (2018) explore why China has failed to develop comprehensive ballast water management policies to prevent invasive aquatic species, despite annually discharging millions of tons of ballast water into Chinese waters, threatening marine biodiversity and causing billions in economic losses. Using Kingdon's multi-stream policy model, the researchers analyzed the problem, policy, and politics streams, illustrating that widespread fragmented governance, a lack of public awareness from poor media coverage, and conflicting institutional priorities prevent effective policy development. Their analysis explains that while China's shipping industry complies with international ballast water regulations when operating in foreign waters, the absence of domestic legislation and bureaucratic entrepreneurs leaves Chinese waters vulnerable to biological invasions from untreated ballast water discharge.

Oberlack and Eisenack (2018) conduct a meta-analysis of 26 global case studies to identify barriers to adapting river basin governance because of climate change. The authors introduce an archetype framework, categorizing identified barriers into patterns, such as coordination gaps, path dependencies, competing priorities, and more. The results emphasize how adaptation barriers are also caused by the collective action of pre-existing institutional arrangements, challenging the conventional focus on biophysical features.

Chen et al. (2021) analyze water security in Xiamen, a rapidly urbanizing coastal city in China that depends on freshwater resources from the Jiulong River watershed. The findings reveal that climate change is exacerbating water management challenges by increasing the frequency of extreme weather events, resulting in more intense flood-drought cycles that directly affect water availability and quality.

Moreover, the increased eutrophication and harmful algal blooms in the Jiangdong Reservoir, the primary water source for Xiamen, are now failing the reservoir's national water quality standards due to anthropogenic activities. By proposing an innovative "source-to-tap" integrated water management framework, the authors emphasize that monitoring the water sources could alleviate this issue.

Horne and Grafton (2019) follow the development of Australia's Murray-Darling Basin water markets, tracking the evolving incremental policy transformations from 1994 to 2018. They explain how the Millennium Drought and coordinated federal-state reform initiatives triggered this evolution, showing how Australia overcame institutional barriers like state rivalries over water resources. Furthermore, the authors highlight and document factors of the political economy that helped transform the water markets, which details how complex institutional arrangements can be successful.

Pahl-Wostl et al. (2023) introduce a framework to analyze "social-ecological fit" in water governance by understanding how coordination structures align with ecosystem service interdependencies. The concept of social-ecological fit is used to compare networks of social interactions with networks of ecological interdependencies, identifying coordination gaps that lead to unsustainable resource management. The findings reveal that while high social-ecological fit in coordination processes can improve water governance outcomes, the transition from improved coordination to environmental sustainability requires multi-pronged governance approaches and long-term transformational change instead of short-term patches.

3.3.2   Institutional Performance in Irrigation Water Allocation

The irrigation sector receives significant attention in the reviewed literature, with studies highlighting both its institutional complexity and interaction with other sectors. Kajisa and Dong (2017) previously mentioned, investigate how volumetric pricing of irrigation water affects water institutions of farm-based "water user groups" (WUGs) and the farmers' conservation efforts.

Ghimire and Griffin (2014) evaluate the distributional and economic impacts of voluntary water transfers from agricultural to urban sectors in Texas, specifically focusing on their irrigation districts' institutions. Using a computable general equilibrium model, the study indicates that water transfers can improve efficiency by reallocating water to high-value uses like the industrial sector, but they create disproportionate losses for the rural agricultural economy. The authors highlight that irrigation districts function as logistical delivery mechanisms and complex sectoral institutions within local economic and governance systems, balancing efficiency and equity within decentralized and locally governed irrigation institutions. In comparison, the municipal and industrial sectors have centralized and regulated institutional frameworks, which complicate coordination and future policies for equitable water transfers.

Venot and Suhardiman (2014) study the constant failure of irrigation sector reforms in case study countries like Ghana and Indonesia by challenging the institutional narrative that deems irrigation as inherently ungovernable because of its complexity. The authors argue that this narrative framing allows agencies and governments to maintain elitist reform models that conceal uneven power dynamics that actually shape irrigation governance. By emphasizing that irrigation institutions are both technically and politically fragmented, we can understand how a variety of workers in engineering, agriculture, and related fields pursue their own priorities and agendas, undermining coherent reform. Although the article mainly focuses on the irrigation sector, it compares the decentralized governance of the residential and industrial sectors with vertically integrated institutions. Rethinking irrigation governance is both a technical implementation and a biased political process issue.

Hoanh and Suhardiman (2014) analyze the evolution of irrigation policy in Vietnam's Mekong River Delta by focusing on irrigation as a sectoral institution. By understanding their irrigation infrastructure and governance, we can see how they have been shaped by the state's priority to increase rice production. Furthermore, they follow a timeline that depicts the shift from centralized planning to fragmented polycentric governance with power imbalances, involving a multitude of actors, such as researchers and governments, in influencing water resource decisions. While the authors emphasize the importance of combining irrigation policy with Vietnam's socioeconomic goals, like crop diversification, they also contrast irrigation development with other sectoral concerns from residential and industrial planning.

Wegerich et al. (2014) study the evolving dynamics of water governance, specifically in the irrigation sector. They argue that the traditional centralized control of irrigation has created fragmented and ineffective governance mechanisms, as evidenced by the emergence of irrigation management transfer, grassroots organizations, and NGOs. Despite efforts to decentralize governance, the authors highlight that irrigation systems are managed by a type of institutional anarchy with weak enforcement and undefined roles, which minimizes effective water governance.

Heinmiller (2016) analyzes how the Great Lakes Basin institutional framework manages irrigation, which tends to supersede the priorities of industrial and residential water use. The study emphasizes the unique governance challenges with irrigation, which have grown more complex because of climate variability and shifting agricultural needs. By assessing transboundary governance through institutional indicators like resilience and legitimacy, it shows how the 2008 *Great Lakes-St. Lawrence River Basin Water Resources Compact* provides a strong framework to coordinate cross-jurisdictional work. Especially since irrigation is considered in a broader sectoral context, we can observe the differences in institutional effectiveness of other sectors' water use.

Saleth et al. (2016) examine the global irrigation sector by focusing on the institutional and infrastructural mechanisms required to manage growing agricultural water scarcity. Although irrigation is an important sector, it has inefficiencies in its water use that could be addressed through demand-side management to increase productivity. The authors describe a few options, such as water pricing and water rights, that could be effective depending on their integration with general legal, policy, and infrastructural conditions. Although the article mainly focuses on irrigation, it helps highlight the intersectoral competition for water. As a result, they argue for a strategically and institutionally integrated approach to water governance by using crises and structural changes as opportunities for agricultural reform.

Kibaroglu (2020) explores Turkey's evolving irrigation governance by studying the initial decentralization efforts through the establishment of irrigation associations, with the aim of improving efficiency and cost recovery. Despite some success, irrigation efficiency did not improve, while local powers slowed participatory performance. The article also contrasts irrigation governance with other sectors' privatization models, demonstrating how public-private partnerships have failed because of commercial risks and a lack of private sector interest.

Villamayor-Tomas (2018) investigates how Spanish irrigation associations respond to disturbances. Using transaction costs and collective action theories, the study categorizes these disturbances by features like intensity and frequency, distinguishing between coordination and cooperation responses depending on the amount of enforcement and collective action needed. Drawing on qualitative comparative analysis of 50 response cases from 5 Water User Associations, the study reveals that cooperative responses are likely when disturbances are internal and progressive, or intense and frequent. He concludes that institutional design and trusting relationships must be enacted, enabling effective decentralized responses to climate and socioeconomic disturbances in the irrigation sector.

Lam (2006) evaluates how Taiwanese irrigation institutions have achieved resilience despite economic and political shifts. By conceptualizing irrigation as a social-ecological system, the study highlights how institutions facilitate coordination among farmers, build adaptive knowledge systems, and manage water among different governance groups. The article focuses on the irrigation sector and its resource allocation, but emphasizes agriculture's political significance and its role in food security despite its economic decline. By tracing the historical evolution of these institutions, we can understand how colonial legacies, state policies, and local farmer engagement strongly influence governance frameworks together.

Janssen and Anderies (2013) use a multi-method framework to analyze the strength of small-scale irrigation systems. Shivakoti and Bastakoti (2006) analyze the strength of Montane irrigation systems in Thailand. Tiger et al. (2011) studied the implications of residential irrigation metering on water demand and consumer expenditures in North Carolina.

Washington-Ottombre and Evans (2019) investigate how the length of tenure among appropriators influences the institutional success of the Mwea Irrigation Scheme (MIS) in Kenya. Using a mixed-methods approach, the study researches operations, maintenance fee payments, and farmers' perceptions of rules governing land, water, and membership. They discover that while newer members are satisfied with current regulations and are less engaged in reform, long-tenured members contribute to stronger institutions by consistently paying fees and actively pressuring for institutional adaptations. They conclude by emphasizing that irrigation is an essential sector to address food security in semi-arid regions, and that the relationship between workers and institutional advancements is vital in managing water resources.

Cody (2019) explores how internalized cultural norms influence performance and the governance of irrigation systems in the Upper Rio Grande Basin. Hunt (2007) examines the social organization of irrigation systems, specifically the communal irrigation institutions. He first distinguishes the three institutional forms of communal irrigation, which are irrigation communities, municipal systems, and special-purpose districts, emphasizing that adequate irrigation is both a technical and a complex, socially organized issue. Moreover, the author critiques the system's dichotomous thinking, such as communal versus bureaucratic, revealing a variety of institutional arrangements from global contexts. His research stresses the importance of clear frameworks for comparative analysis and further argues that communal irrigation systems tend to be more sustainable and politically resilient than state-run systems.

Gutiérrez-Malaxechebarría (2013) researches the widespread phenomenon of informal irrigation in Colombia's Andean regions by analyzing how small-scale farmers independently develop, manage, and adapt irrigation systems outside the state's formal oversight. They often use local knowledge, collective agreements, and simple infrastructure, such as hoses and sprinklers. Using a mixed-methods approach, they map the variety and usage of informal irrigation systems and their role in maintaining household agriculture. Emphasizing the tensions between formal state institutions and local management, the author proposes a hybrid governance model that uses informal practices while promoting sustainable and equitable water access across all sectors.

Behera and Mishra (2018), evaluate the performance of the long-known water management institution in Panchayat, India.

Pradhan and Ranjan (2015) study how institutional interventions aimed at groundwater augmentation influence farmers' crop choices in drought-prone regions of Andhra Pradesh, India. Using a Multivariate Probit model, the results show that while programs, such as the crop water budgeting (CWB), were designed to promote sustainable water use, they ironically increase the likelihood of farmers cultivating high water-intensive crops like rice. This exposes the mismatch between policy intentions and actual outcomes, suggesting that well-off farmers may be utilizing these programs to mitigate short-term risks instead of prioritizing long-term groundwater conservation. By embedding water extraction compliance rules into institutional support mechanisms, the authors argue that it can prevent contradictory incentives.

Kimmich and Villamayor-Tomas (2019) use a unique approach to analyzing resource governance by examining networks of "action situations" (AS's) within irrigation systems, specifically analyzing Spain and India. Using configurational analysis and centrality measures, the authors study how interactions between water and energy governance structures affect the outcomes. By comparing the drought-resilient Spanish surface irrigation system and India's inefficient groundwater system, the authors discuss the dynamic relationships between institutional arrangements, actor strategies, and technical infrastructure. They specifically integrate the network theory and game-theory modeling to identify which governance situations create the intended outcomes, so just improving infrastructure or technology will fail if the underlying coordination and institutional dilemmas are not addressed. Hayden and Tsvetanov (2019) investigate irrigation restrictions in California and the effectiveness of the outdoor irrigation water restrictions.

Van der Kooij et al. (2015) analyze institutional transformation within the irrigation sector through a socio-technical lens, focusing on the Sguia Khrichfa canal in Morocco managed by farmers. The authors argue that irrigation institutions are shaped through the evolution of social and technological contexts, like the creation of drip irrigation. Rather than understanding technology as just a tool, the study emphasizes how it actively controls authority, access, and distributional arrangements of a variety of irrigators.

Nüsser et al. (2019) investigate the growing vulnerability of cryosphere-fed irrigation systems relying on snow and glacier melt, especially considering the impact of climate change. The authors recognize that many irrigation systems in the Northwestern Himalaya were designed explicitly around predictable snowmelt timing, but are becoming increasingly unaligned with changing runoff patterns, threatening water availability and institutional resilience. Arguing that the irrigation sector lacks adaptive capacity, the authors believe that it cannot rebound like the residential or industrial water sectors, which often have infrastructural flexibility, financial capacity, and more centralized command. They further critique the narrative that irrigation failures are technical or managerial when they should be framed as institutional challenges created by outdated knowledge about hydrological regularity. By proposing a reconfiguration of irrigation governance, the authors believe we can account for the long-term restructuring of water regimes as climate change worsens.

Mottaleb et al. (2019) understand how institutional arrangements shape irrigation service delivery in rural Bangladesh. Focusing on the irrigation sector, they study how water is accessed through informal markets driven by smaller private providers, and how different water contract types are created based on local social norms, market competition, and provider-client relationships. Through fieldwork and survey data, the authors illustrate that institutional structures are critical in the efficiency and equity of irrigation services. Ultimately, the authors suggest integrated and context-sensitive water governance as a solution.

Sakketa (2018) studies how communal irrigation systems are organized and maintained through institutional bricolage, which is defined as the adaptive recombination of formal rules, informal rules, and norms. He specifically highlights how the irrigation sector's systems operate based on the intersection of state policies, community governance, and local traditions, creating hybrid forms of water management.

Zhang et al. (2014) analyze how evolving market conditions influence the local irrigation institutional frameworks that govern water management in Northern China. By investigating how irrigation institutions adapt in response to external pressures like commercialization and market integration, the authors emphasize how formal water-user associations coexist and interact with informal practices, which reshape access rules, enforcement, and collective decision-making. Tradable water use rights were introduced to improve the efficiency of water allocation; however, formal water trading is still not used as much because informal arrangements are used instead due to high transactional costs, limited awareness, and strong interpersonal relationships.

Xie and Zilberman (2016) develop a formal analytical model to examine how institutional rules, environmental variability, and operational conditions influence decisions about the sizing of irrigation water infrastructure. The irrigation sector's governance mechanisms, like allocation rules, heavily impact the capacity of water projects. However, by integrating institutional dynamics into technical planning, the model provides a framework for pairing infrastructure design with the challenges of water governance.

Wang et al. (2013) investigates how physical attributes of natural resources, governance structures, and institutional rules collectively shape irrigation management for villages in Northern China. The authors compare two villages with different surface and groundwater systems to illustrate how local agents adjust rule structures in response to environmental constraints or social organizations. While groundwater systems cultivate bottom-up and trust-based management, the results indicate that surface water systems rely on contractor-based governance with a hierarchy model.

Yang et al. (2003) examine China's experimental reforms in water rights and markets, specifically in the irrigation sector as well. They discuss how pilot programs in various provinces have attempted to clarify the use of water rights, introduce transferable quotas, and incentivize conservation-based behaviors for agricultural users. The study emphasizes how institutional design influences the outcomes of water market reforms, exemplifying how adjusting these designs can ensure equitable and effective water governance.

Li et al. (2017) construct an economic model to analyze how farmers' decisions on irrigated land use are influenced by uncertainty in water supply, with a focus on systems that allocate based on priority of water rights. The results show that farmers with higher-priority water rights are more likely to expand their irrigation because of reduced exposure to supply risk. By combining risk aversion and water entitlement frameworks, readers can better understand how formal institutional designs impact environmental uncertainty through land use and resource efficiency to shape adaptive behavior.

Olen et al. (2016) study how water scarcity and climatic factors influence farmers' decisions across major crop types on the West Coast of the United States. By analyzing cross-sectional data to assess changes in irrigation behavior in crops like almonds and grapes, the results illustrate that both long-term climatic determinants and short-term drought conditions considerably impact the frequency and intensity of irrigation. The authors stress how institutional constraints, such as water availability and accessibility, and rapidly changing environmental conditions, control farmers' decision-making process.

Li and Zhao (2018) investigate how utilizing water-efficient irrigation technologies can unintentionally cause increased water use, a phenomenon also known as the rebound effect. Researching the Ogallala-High Plains Aquifer region's irrigation, the authors conclude that water rights can amplify this rebound effect because farmers with larger entitlements usually irrigate more land to grow more water-intensive crops. The results display that the rebound effect is half-and-half caused by land expansion and intensive irrigation. Through an empirical model, the authors show how water rights can control the environmental outcomes of technological interventions, which can be solved by creating reforms that take into account different interventions.

Dridi and Khanna (2005) create a theoretical framework analyzing how asymmetric information between farmers and regulators influences water allocation, adoption of irrigation technology, and water trading. They further elaborate on how incomplete details on farmers' land quality reduce incentives to adopt efficient irrigation technologies because of discrepancies in water quotas and pricing mechanisms. Introducing bilateral water trading, the authors describe how it can enhance social welfare through improved resource allocation and encourage the adoption of irrigation technologies, even with asymmetric information.

Kahil et al. (2015) evaluate the water market and irrigation subsidy policy approaches to encourage irrigation adaptation to climate change in Southern Europe. As climate change grows, it will negatively affect irrigation and water-dependent ecosystems, but the extent of these impacts depends on government policy and farmers' adaptive strategies. The authors conclude that market-based trading tends to generate better economic outcomes compared to technological subsidies through the promotion of efficient water distribution. However, they also warn of tradeoffs where markets could reduce ecological water availability, negatively impacting river flows and groundwater reserves.

Michalak (2020) investigates the effectiveness of water management policies in Polish agriculture, addressing how farms and governmental institutions adapt to climate change. He uses a dual-level analysis to assess institutional support from governmental entities and farmers' behavior through case studies, aiming to understand how Polish farms manage water resources during periods of water scarcity as climate change worsens. Despite minimal institutional assistance, larger agricultural enterprises independently adopt irrigation technologies and drought-resistant farming practices. It shows how farmers are unaware of how their water management decisions affect ecological systems and the economy.

Suarez Bosa (2015) examines the unique water management institutional framework in Cape Verde, which combines both public and private governance structures for irrigation water resources. The study reveals a dual system where water from natural springs operates under private ownership with individualistic management practices, whereas water from wells functions under community-based management despite state ownership. Through qualitative surveys and interviews, the author explains that community-managed water systems are more efficient than individual management approaches, which tend to waste resources. Interestingly, the institutional model engrains a blend of Portuguese colonial legacy and African cultural traditions, which was a mutualistic relationship that facilitated cooperative water distribution.

Nyagumbo and Rurinda (2012), evaluate the effectiveness of various water management institutions for smallholder farmers in Zimbabwe.

Iglesias et al. (2007) introduce their "Economic Drought Management Index" (EDMI), which is designed to analyze the efficiency of water allocation decisions in an unstable water environment. Based on a stochastic dynamic optimization model, the EDMI evaluates the trade-off between immediate water use and future water security, integrating key variables such as reservoir storage capacities, institutional decision rules, and the economic value of irrigation water. The authors empirically apply this index to two irrigation systems in Spain's Guadalquivir River Basin, in which one is centrally managed and one is self-managed, displaying that the self-managed system performs more efficiently in mitigating drought risk.

Meinzen-Dick (2007), examines the experience of searching for panaceas in the irrigation sector.

Jyotishi and Rout (2005) investigate water rights and resource management in India's Deccan region by analyzing the Baliraja movement and other institutions. The Baliraja movement maintained an idealistic approach of equitable water distribution, and they believed in equal water rights for all households regardless of the landholding size. By comparing the Baliraja movement and the more practical and cooperative water institutions that thrived, the results indicate the movement's failure. While it aimed to address equity, efficiency, and sustainability through restricted crop choices and equal water allocation, the movement ultimately failed because it limited farmers' economic opportunities and crop selection, especially with sugarcane. In contrast, the other successful water institutions prioritized water supply for high-value crops like sugarcane, creating practical arrangements despite inequalities. The authors conclude that policymakers must choose between equity-sustainability or equity-efficiency solutions.

Zhang and Oki (2023) study China's nationwide agricultural water pricing reform for sustainable water resource management in developing countries. The research reveals that China has implemented an approach that combined "reasonable pricing" mechanisms with "precise subsidies and water-saving incentives," including investing in infrastructure and quota control management systems. China uniquely focused on farmers' affordability, targeting operation and maintenance (O&M) cost recovery to eventually progress toward full water supply cost pricing. The results display substantial water conservation achievements, even with varying progress across China's provinces, emphasizing regional disparities in economic development and infrastructure capacity. The authors explain how China balances food security concerns with sustainable water management by focusing on comprehensive policy integration instead of just relying on price adjustments to achieve conservation.

Seijger and Hellegers (2023) introduce the concept of "reorientation" to analyze how societies reform their agricultural water management systems in response to changing societal priorities about water, agriculture, and environmental concerns. The authors define reorientation as "a shift in broader societal priorities that drives reform of agricultural water management," setting this phenomenon as a social process that occurs over decades to bridge the gap between agricultural change and strategy-specific optimization. By examining 21 global reorientation examples, the authors indicate the diversity in starting points of priorities, the extent of successful implementation, and the role of government intervention. The authors identify that reorientations toward agricultural expansion and intensification appear to be more easily accommodated than those for environmental conservation, suggesting that social and biophysical boundaries limit agricultural water management reformations.

Iyiola et al. (2024) examine the role of sustainable water use and management to achieve agricultural transformation across Africa, emphasizing the interwoven challenges of water security, food production, and economic development. The authors describe agrarian transformation as a gradual process that transforms subsistence-based farming to a productive and commercialized agricultural system, identifying three pathways for this transformation to drive economic growth. This includes multiplier effects on relevant agriculture sectors, factor market impacts, and increased household consumption of agricultural products. Moreover, this study undertakes a unique and holistic approach to identifying specific measures for agricultural transformation in Africa, including improving agricultural value chains, enhancing institutional capacity, and incorporating youth and women. However, the authors reveal significant challenges unique to the African context, such as the continent's limited usage of its agricultural potential despite owning many fertile lands, vast water resources, and limited irrigation.

3.3.3   Institutional Performance in Urban Water Allocation

Urban water institutions are addressed in various studies focusing on indicators, governance challenges, and institutional evolution. Polonenko et al. (2020) analyze institutional and social indicators used in understanding urban water systems, emphasizing the importance of these indicators for evaluating sustainability and resilience. Although technical and economic measures are well established, the effective use and implementation of social and institutional indicators are ambiguous. They develop a framework to better use these indicators efficiently by focusing on criteria, like reliability, validity, practicality, and generalizability.

Similarly, Byrnes (2013) uses the institutional and regulatory history of the urban water sector in Australia to improve water efficiency using the same indicators. He traces the sector's four key phases of evolution to identify regulatory fragmentation, corporatization, and reforms for efficiency. Furthermore, Kayaga et al. (2013) developed another model to examine the infrastructure of these institutions through the water utility maturity (WUM) model. They use the indicators above to break down the capabilities of institutional capacities, so they highlight the balance between capacity and performance for the best results and sustainability of the urban sector.

Punjabi and Johnson (2019) dissect the complex institutional dynamics of rural and urban water conflicts in India to understand how water governance shapes resource allocation between sectors. By comparing Mumbai and Chennai, they observe how Mumbai uses a historical framework of prior appropriation, allowing urban authorities to control rural water resources, which disadvantages tribal irrigation communities. However, Chennai allocates water depending on the market and riparian rights-based institutions, leading to the commercialization of groundwater, resulting in unsustainable extraction practices.

Fuenfschilling and Truffer (2016) examine the transformation of the Australian urban water sector and its governance using a socio-technical lens, understanding how institutional structures, decision-making, and technology interact and adapt during environmental crises. Using the concept of "institutional work", the authors compare and contrast how desalination was quickly institutionalized despite high costs and sustainability concerns, but wastewater recycling was opposed. Similar to Punjabi and Johnson (2019), we can observe how power dynamics in the urban sector are at play, influencing how technological pathways and infrastructure systems are shaped.

Bettini et al. (2015) analyze governance through the concept of a "scale", challenging the conventional view that it is purely a technical matter. Similarly, Bakker et al. (2008) analyze Jakarta's urban water supply systems by understanding failures in governance, highlighting how both institutional structures and ownership provide access to water for poor households. The results demonstrate that Jakarta's water infrastructure is a fragmented system where decisions are made by political agendas and exclusionary planning, prioritizing higher-income and industrial zones. The authors conclude that institutional reform is more important than just private or public ownership to achieve equitable access to water.

Scott and Pablos (2011) study the evolving governance of urban water and its wastewater management in Hermosillo, New Mexico. They describe how cities transform into institutional innovations by using their urban hydraulic reach during periods of water scarcity, which is the process of reallocating and redistributing rural water sources and wastewater. In the Hermilloso case study, the authors argue that governance was run by negotiated arrangements between actors rather than centralized or market-driven reforms.

Brodnik et al. (2017) investigate the institutional dynamics, stability, and practice change in Australia's urban water management sector from 1970 to 2015. The authors identified six distinct institutional logics, which are decision making, risk, sustainability, water quality, infrastructure, and demand, that evolved, explaining both sectoral stability and the emergence of sustainable urban water management practices. The findings challenge existing institutional theory by demonstrating that practice change results from incremental co-evolution of multiple institutional logics instead of replacement or competition. They conclude that sustainable practices like water recycling and stormwater harvesting emerged through coordinated changes across multiple levels instead of one institutional framework.

Thomas (1995) focuses on industrial water use and investigates how regulatory frameworks target pollution control despite information being asymmetrical between regulators and industrial polluters regarding waste generation. Results find that the imperfect pollution tax or other similar policies did not effectively impact pollution levels, so they suggest implementing a "contract-based regulatory scheme". The authors also emphasize how the pollution standards and effectiveness depend on the industries, like chemical industries, which are more efficient than food manufacturers.

Eisenack (2016) analyzes institutional adaptation to water scarcity for thermo-electric power generation under climate change, using a mathematical model to compare the three regulatory arrangements based on transaction costs and environmental externalities in the German Rhine catchment. The author develops a qualitative reasoning model that analyzes how the frequency and intensity of heat waves influence the comparative costs of different institutional arrangements, finding that temperature caps perform best when heat waves increase only in intensity, while more complex arrangements become cost-effective when heat wave frequency rises. The results explain that the introduction of Germany's minimum power plant concept after the 2003 heat wave aligns with the model's predictions. This suggests that institutional path dependency will most likely prevent further changes despite ongoing climate change, especially since Germany is transitioning away from thermo-electric power generation.

Saleth (2018) studies the institutional economics of water by reviewing nine papers that analyze how water institutions create incentive environments for sustainable water management across various regional and sectoral scales. The papers use a variety of methodological approaches, ranging from econometric analysis and game theory to historical analysis and case studies, explaining topics like water governance, transaction costs of water transfers, and the evolution of water supply institutional structures. The findings display that water governance operates through economic calculus, so institutional performance depends on the relationships between institutional, economic, and political branches. The author also highlights how institutional inconsistency can create contradictory effects and limit the effectiveness of water management organizations across different contexts.

Keen (2003) examines the challenges facing integrated water management in the South Pacific, using Fiji's capital city named Suva, as a case study to illustrate the multifaceted nature of water governance issues in Pacific Island countries. Uniquely, she focuses on the socio-cultural barriers to water management reform in Pacific Island contexts, highlighting how customary land tenure systems and traditional concepts like "vanua", which equates land, water, and human habitats as the same, create distinctive governance challenges not found anywhere else. Although the South Pacific faces similar technical water infrastructure problems as other developing nations, their small scale, unique geographical characteristics, and cultural contexts require fundamentally different institutional approaches to water management. The author argues that effective integrated water management must address technical capacities, infrastructure development, and the complex relationship between traditional governance systems and modern regulatory frameworks, emphasizing the need for culturally sensitive policy approaches and inter-agency coordination to achieve sustainable management.

Ferguson et al. (2013) analyze Melbourne's transformation from traditional centralized water infrastructure to an integrated hybrid system between 1997 and 2012, mainly driven by the Millennium Drought that created significant institutional changes across cultural-cognitive, normative, and regulative dimensions. Their research identifies key causal factors, such as shifts in professional beliefs about environmental predictability, development of new technical knowledge, expanded water servicing goals, and improved governance coordination across previously excluded sectors. The authors argue that successful water system transformation requires technological innovation and comprehensive institutional reform.

Shah and van Koppen (2006) study on whether the institution of IWRM is fit for India asks five critical questions regarding the IWRM paradigm and India that should be considered by all countries when investing in whether IWRM is an appropriate institution to use. The questions include "(1) Is water poverty in countries caused by their water scarcity? (2) Would embracing IWRM help alleviate India's water poverty? (3) Is implementing IWRM feasible in India in today's context? (4) Has implementing IWRM helped counter water scarcity and poverty in other countries with a development context comparable to India's? Furthermore, finally, (5) What should be the priorities and roadmap for improving the working of the water sector in India? They conclude that water pricing and withdrawal permits have been challenging to implement administratively, and renaming territorial entities into river basin organizations has failed to manage river basin water. They argue that while the IWRM direct demand management model is not incorrect, it is impractical in informal water economies. The rise of a class of intermediaries between users and natural water sources, in the shape of water service providers, is a precondition for meaningful demand management. The emergence of a new class of middlemen between users and natural water sources, known as water service providers, is a prerequisite for effective demand control.

Satoh (2019) examines the effectiveness of water rights. Muller (2014) examines South Africa's comprehensive approach to creating constitutional rights to water and sanitation following the end of apartheid in 1994. The author analyzes the policy and legislative framework that granted a vast majority of households water by 2011, representing one of the most successful large-scale water service delivery programs in the developing world. The case study had a variety of implementation strategies, such as the agenda that simultaneously prioritized basic access and improving service quality, and the innovative "free basic water" policy. The findings also identify critical success factors, such as strong political multi-level commitment, substantial national budget allocations, and the use of technical infrastructure. However, the author describes ongoing challenges like the inadequate operation and maintenance of infrastructure.

Kremer et al. (2011) evaluate spring protection interventions in rural Kenya, focusing on the health and economic implications of water infrastructure. The authors find that spring protection "reduces source water fecal contamination by 66%" and "decreases child diarrhea by 25%". Furthermore, households' willingness to pay for these improvements was significantly lower than standard health predictions, suggesting highly income-elastic demand for preventive health measures. The results also explain that current communal property norms, although limiting private investment incentives, may actually increase social welfare more than privatized regimes because of the inefficiencies of above-marginal-cost pricing.

Nauges and Whittington (2019) evaluate social norms information treatments (SNITs) in municipal water supply systems, addressing welfare implications. The authors develop and present a four-piece framework that considers implementation costs, utility cost savings, household welfare effects, and environmental benefits, revealing that SNITs may widely fail social benefit-cost tests because of modest treatment effects and potential costs imposed on households. While SNITs and price increases achieve similar short-term water conservation, price increases generate significantly higher net social benefits, especially in low and middle-income countries where water is usually priced below cost. The authors challenge SNITs by displaying that they should only be used as temporary drought management tools instead of permanent conservation policies, arguing that it takes away from politically challenging tariff reforms.

Van De Meene and Brown (2009) dissect the myth behind the transition to sustainable urban water management, providing institutional attributes required for this transition. Conducting a comprehensive literature review of 81 studies, the authors develop a four-sphere framework, which includes administrative/regulatory, inter-organizational, intra-organizational, and human resources components. The results reveal the differences between traditionally hierarchical, centralized water management systems and the proposed sustainable governance. The authors argue that achieving sustainable urban water management requires a fundamental paradigm shift from command-and-control governance toward network-based approaches emphasizing learning and collaboration. The analysis suggests that technical solutions and changes in governance structures must be made simultaneously.

Means et al. (2002) warn readers about a water infrastructure crisis in the United States, tracking the drastic increases in estimated replacement costs from almost $138.4 billion in 1997 to nearly $1 trillion by 2000. After decades of declining infrastructure investment, the government has created a maintenance deficit that risks doubling or tripling water rates, with smaller systems being disproportionately impacted due to a limited rate base. By analyzing the political ramifications of rising water costs, the authors predict increased "politicization" of water management because candidates may run campaigns on rate reliefs or open the doors to privatization.

Krishnaraj (2011) analyzes water governance in India, understanding how intersecting inequalities of gender, caste, and class influence access to and control over water resources, even with policies encouraging women. The results display how decentralization policies that should empower women to manage domestic water supplies actually reinforce traditional hierarchies, bearing increased responsibilities without the appropriate knowledge. The author details different historical exclusions, such as dalits being banned from village wells, arguing they still persist in modern community-based water management despite aims for equitable traditional water governance. Interestingly, progressive participatory approaches can paradoxically fortify existing power structures by failing to address underlying social inequalities, showing similarities to Li and Zhao's (2018) introduction of the "rebound effect", another paradoxical increase.

Rogers et al. (2015) provide an institutional analysis of water innovations in Melbourne, researching how various institutional works facilitate the development of institutional innovations. The three different institutional works they examine are cultural-cognitive, normative, and regulative. By looking at case studies of desalination, wastewater recycling, and stormwater harvesting, the authors explain how successful innovations only occur with institutional alignment in established governments. Citing Lawrence and Suddaby's institutional work concept and transitions theory, the authors argue that an agency-centric lens creates the institutional foundations necessary for technological change in infrastructure to improve sustainability goals.

Grison et al. (2023) assess Integrated Water Resources Management (IWRM) performance across global cities, specifically examining "200 cities in total [that] represent more than 95% of global urban population". By using the City Blueprint Approach lens to measure urban water management effectiveness, the authors identify 24 performance indicators, which include infrastructure quality, wastewater treatment, climate resilience, and governance frameworks. Uniquely, the authors use an innovative statistical estimation model to conclude that government effectiveness deters urban water management success, supporting the institutional perspective that effective governance forms a successful foundation for management. The findings expose global disparities in water management capacity, highlighting substantial deficiencies in progress toward water-related Sustainable Development Goals (SDG), especially in African and Latin American urban areas that severely lack infrastructure and management.

Martínez-Valderrama et al. (2023) explore the widening gap between agricultural water demand and renewable water supply, demonstrating how traditional supply-side solutions paradoxically exacerbate water scarcity issues they are intended to solve. The authors explain how conventional approaches create self-reinforcing cycles where increased water availability generates greater demand, ultimately leading to resource depletion and ecosystem collapse, as seen in the Aral Sea case study. The findings deconstruct four vital mechanisms driving water gap expansion, which include the "reservoir effect" that discourages demand management, the Jevons Paradox in irrigation efficiency reforms, the growing competition from non-food crops used for biofuels and livestock feed, and the prioritization of mass production over sustainability. The authors conclude that closing the agricultural water gap requires a multi-pronged policy approach tailored to both local and global environments, aiming to move toward comprehensive demand management and sustainable resources.

Ouyang et al. (2024) evaluate China's water resources tax policy pilot program implemented across ten provinces from 2016 to 2019, defining a notable shift from the traditional fee-based system to a stringent tax-based approach for water governance and conservation. By combining theoretical economic analysis and field investigations, the authors assess the effectiveness of the Tax-for-Fee reform to demonstrate how water resources taxation influences consumption patterns through price mechanisms and behavioral changes toward alternative water sources, such as recycled water. The findings identify success factors for water tax implementation, such as stakeholder interest balance and flexible local government discretionary powers.

da Silva et al. (2025) study institutional arrangements for transboundary water governance in the Quaraí sub-basin along the Brazil-Uruguay border, emphasizing the role of rice farmers in shaping cooperative frameworks. The findings demonstrate that while strong social and cultural integration between Brazilian and Uruguayan border communities helped establish informal water management institutions during the rice cultivation expansion in the twentieth century, these informal ties were insufficient in maintaining adequate formal institutional arrangements. The authors illustrate how local water disputes, especially during drought periods, actually triggered the development of stronger transboundary governance mechanisms, such as the "Gauge Rule" system that regulates water extraction based on river gauge levels. The authors argue that the Quaraí basin case study exposes the tensions between local and cross-border cooperation and national bureaucratic priorities, in which formal transboundary institutions remain vulnerable to political neglect despite successful grassroots collaboration.

3.3.3.1   Water Management

Huang et al. (2009) examine the institutional transformation of water management systems in northern China between 1995 and 2004, revealing that some villages shifted from traditional collective management toward market-oriented reforms, including water user associations (WUAs) and contracting arrangements. Counterintuitively, villages with relatively abundant water resources and complex irrigation infrastructure were more likely to implement reforms than those experiencing acute water scarcity, suggesting that institutional change occurs under operational flexibility rather than necessity from crisis. While contracting systems effectively utilized financial incentives for water conservation, WUAs struggled to achieve effective participatory governance because village leadership maintained control, highlighting the gap between goals for reform and results from implementation.

Similarly, Huang et al. (2010) present an empirical analysis of institutional reforms in Northern China, focusing on the replacement of traditional collective water management through water user associations and contracting systems. Drawing on data across three provinces, the results reveal that water user associations, intended to be participatory, are actually controlled by existing village leadership, limiting farmers' participation. However, the limited democratization outperforms the traditional systems, in which the performance can be measured by improved metrics in fee collection and timely water delivery.

Ferguson et al. (2013) analyze Melbourne's transformation from traditional centralized water infrastructure to an integrated hybrid system between 1997 and 2012, mainly driven by the Millennium Drought that created significant institutional changes across cultural-cognitive, normative, and regulative dimensions. Their research identifies key causal factors, such as shifts in professional beliefs about environmental predictability, development of new technical knowledge, expanded water servicing goals, and improved governance coordination across previously excluded sectors. The authors argue that successful water system transformation requires technological innovation and comprehensive institutional reform.

Easter and McCann (2010) explain how there has been overinvestment in water infrastructure and underinvestment in water institutions historically. They alternatively propose the need to place a higher focus on the creation of new and better water institutions, particularly with water rights.

3.3.3.2 Incentives

Muchapondwa et al. (2018) aim to identify successful market-based incentives to help promote sustainable watershed practices.

Rai et al. (2019) studied water availability in a small town in Nepal, identifying the ideal institutional framework for implementing an incentive payment for ecosystem services (IPES) and a fund flow mechanism.

Zuo et al. (2016) developed innovative methods to measure price elasticities for water entitlements in Australia's Murray-Darling Basin, the most developed water market system. They combine preference survey data with preference market transaction data, which overcomes the traditional restrictions of data and endogeneity issues in elasticity studies. By surveying irrigators using a contingent behavior approach, the authors find that both "demand and supply of high security water entitlements are relatively inelastic within the relevant market price range of AUD $1700–$2100 per megaliter, with supply being more inelastic than demand". The results demonstrate that irrigators are generally unresponsive to price changes in their water trading decisions and tend to overestimate their willingness to sell water. As a result, the authors argue that limited price responsiveness suggests that traditional market-based approaches may struggle to reach water reallocation goals, necessitating alternative policy solutions.

Cobourn (2015) studies the externalities and simultaneity in surface water-groundwater systems and challenges for water rights institutions, examines how water-use decisions and improvements in irrigation technologies affect communication across space and time, posing a challenge for policy instrument design.

Gharib et al. (2024) developed a methodology that evaluates how population location under alternative water institutions and climate scenarios impacts water demands, shortages, and derived economic values.

Lefebvre et al. (2012) explore whether security-differentiated water rights or single security systems improve water market performance. The authors use an experimental design with different transaction cost arrangements for a water rights market versus a water allocation market, testing how security differentiation affects allocative efficiency and risk management. When transaction costs are higher in the water allocation market than the water rights market, the findings demonstrate that different water rights increase overall profits and improve market efficiency, suggesting that the cost structure between markets is crucial.

3.4 Multi-sector Institutions, Economy-Wide Institutions (Water-Energy-Food)

Fifty-six papers described the different types of multi-sector institutions or institutions considered economy-wide. Multi-sector water institutions are considered to be impactful, economy-wide, not only in the water sector but also in the energy and food sectors. According to a recent study by Hoekstra et al. (2019), food, energy, and numerous manufactured commodities depend on the reliable availability of sufficient and clean water.

Regarding economy-wide institutions, the most important aspect to consider in relation to water institutions is water security. Grey and Sadoff (2007) define water security as the accessibility of an adequate amount and nature of water for human wellbeing, livelihoods, biological systems, and efficiency, as well as a low degree of water-related dangers to individuals, the climate, environment, and the economy.

Water is increasingly viewed as a critical global public good that should be administered through a global, interconnected, and participatory strategy rather than a separate sector. The connection between water and major global concerns underscores the importance of global water cooperation in establishing a broad, multi-sectoral agenda. One of the most important institutions directly associated with multi-sectoral water use is Integrated Water Resources Management (IWRM). One comparative cross-country study of 13 countries (regions) discusses IWRM from the perspectives of the Rio-Dublin principles, the Global Water Partnership definition, and concepts and ideas from adaptive governance. Essentially, water integration in this study is the centralized core of IWRM, which includes three identified forms: functional, societal, and institutional integration. They conclude that for future use, IWRM requires more systematic and comparative methods from the perspectives of both administrative (de)centralization and economic growth (Lubell and Edelenbos 2013). Similarly, Stoa (2017) study on water governance in Haiti, which also uses IWRM, concluded that water resource management planning should involve local governments and community institutions, with national or international assistance supplementing capacity.

Barbier (2019) focuses on reforming governance and institutions, covering different sets of studies on groundwater and freshwater, and recognizing that "there is a mismatch between water governance and institutions and our management needs." Today, the world is depending more and more on groundwater. Yet, the technological advancements that have made it easier to extract, access, and use more groundwater than ever before have also jeopardized resource management. Barbier (2019) study states that "global abstraction of groundwater has increased fourfold in the past 50 years, and today it provides 36% of potable water, 42% of irrigation for agriculture, and 24% of water for industry. Groundwater now supplies over a billion urban residents, and many rural households throughout the world". The technological advances that have made it easier to abstract and use more groundwater than ever before have also undermined the governance of the resource. Overall, the study finds that the most important challenge for improvement in governance and institutions is to "overcome the chronic underpricing of water resources."

Similarly, Mukherji (2007) study on India's groundwater-sharing institutional arrangements believes that sustainable groundwater institutions that can control unsustainable groundwater extraction rates are a big problem. The duration aspect of the groundwater right is attenuated differently in different institutional arrangements. For example, Ananda and Aheeyar (2020) point out that the kinship group-owned tube wells in West Bengal's water-sharing laws allow each sub-group of farmers to use 12 h of irrigation water every 36 h. The study stated that the Groundwater price regulation implemented by the Panchayat society in the same state did not affect the de facto duration of the groundwater rights.

Frameworks and parameters are an important part of multi-sector institutions that shape how they are governed. Sears et al. (2019) investigate Spatial Groundwater Management in California, using a dynamic game framework for analyzing spatial groundwater management. The study describes the non-cooperative Markov perfect equilibrium and compares it to the socially optimal coordinated solution. They then calibrate the dynamic game framework to California and conduct a numerical analysis to compute the deadweight cost stemming from non-cooperative behavior in order to examine the benefits of internalizing spatial externalities in California. Their results show that higher crop returns, electricity input prices, whether the crop is annual or perennial, the amount of the groundwater pool, the region's climate, and the adjustment costs of fallowing output all contribute to inefficiencies caused by spatial externalities. When crop prices are high in California, the benefits of coordinated management are very significant. Ludwig et al. (2012) address ways that the water sector can adapt to climate change and reduce the negative impacts of flooding, water shortages, drinking water, water for sanitation, water for industry, and water for crop irrigation. Araral (2010) evaluates the joint performance of institutions and infrastructure in impacting the effectiveness and efficiency of the water sector. Empirical examples are drawn from different countries and regions around the world. Gupta and Pahl-Wostl (2013) explore the challenges of global water governance and address whether water challenges are global challenges and what type of governance and water resource management would be appropriate. The article suggests that management at the river basin level is now overtaken by the worldwide water governance.

Grafton et al. (2019) use WGRF to establish water supply and demand convergence to maintain the sustainability of freshwater ecosystem services. The WGRF consists of seven critical strategic considerations regarding water reform and implementation of the framework, which is designed to not only be flexible for use in any country but also be integrative in regard to the reform research goal, inequities in water allocation, and simple to use or flexible to many scales and contexts. The seven considerations of WGRF include: "(1) well-defined and publicly available reform objectives; (2) transparency in decision-making and public access to available data; (3) water valuation of uses and non-uses to assess trade-offs and winners and losers; (4) compensation for the marginalized or mitigation for persons who are disadvantaged by reform; (5) reform oversight and "champions"; (6) capacity to deliver; and (7) resilient decision-making that is both beneficial and durable from a broad socio-economic Perspective". The study applies WGRF to the following countries: the Murray–Darling Basin (Australia), Rufiji Basin (Tanzania), Colorado Basin (shared by the USA and Mexico), and Vietnam. The study's evaluation of WGRF in these five countries proved that the framework was successful. They believe it should be the "core" of water governance reform efforts since it is adaptable enough at the local, basin, and national levels. For the Murray-Darling Basin and Australia, they provide a framework for assessing reform successes and failures, highlighting power imbalances, and, most crucially, providing constructive assistance for adaptive reform management. Second, when applied to Tanzania's Rufiji Basin, the WGRF examines the existing water governance system and, as a result, suggests future water reform options.

Gopalakrishnan et al. (2005) report the results of a global survey and assessment of the structure, evolution, and performance of water institutions at various settings: regional, national, and international. It includes an analysis and discussion of the rationale for institutional innovations in Hawaii, Sri Lanka, India, China, Mexico, and several international comparisons in the Middle East and in Western USA.

Programme (2009) emphasizes that community-based action is critical to address the core reasons of 'leaving people behind' in terms of water and sanitation. Water resource allocation mechanisms can be set up to meet a variety of socioeconomic policy goals, such as ensuring food and/or energy security or promoting industrial growth, but ensuring that enough water is available (and of suitable quality) to meet everyone's basic human needs (for domestic and subsistence purposes) must be a top priority. The UN member states have made commitments to adopt the 2030 Agenda for Sustainable Development and recognize the human rights to safe drinking water and sanitation, both of which are critical for reducing poverty and building successful, peaceful communities. Ghimire and Griffin (2014) combine the institutional and political orientation of irrigation districts, favoring irrigation over irrigators. Tortajada and Joshi (2013) describe the development, by the city-state of Singapore, of a comprehensive plan for the overall management of its water resources.

Middleton and Dore (2015) study on transboundary water and electricity governance in mainland Southeast Asia investigates the links and disjunctions between regional electricity and water governance frameworks and the context, drivers, instruments, and arenas of water and electricity decision making. Linkages between power administration and water administration, while for the most part frail and packed with power imbalances, are distinguished, including Ecological Effect Appraisal and Key Natural Evaluation devices. Disjunctures incorporate state sway and restricted coverage of entertainers among the power and water administration fields.

VanNijnatten et al. (2016) study on assessing adaptive transboundary governance capacity (TGC) in the Great Lakes Basin discusses how transboundary governance began with the 1909 Boundary Waters Treaty (BWT) and the establishment of the International Joint Commission (IJC). In reaction to unused logical data, almost genuine contamination impacts within the Bowl from industrialization and urbanization, the 1972 Great Lakes Water Quality Agreement (GLWQA) put in place additional transboundary instruments to address water contamination. Reestablished in 1978, 1987, and 2012, this "non-official, good-faith understanding between the two levels of government" comes about in a center on 43 Areas of Concern (AOC) and the execution of Remedial Action Plans to clean up and delist contaminated waters. A collaborative foundation was built up to control chemical inputs and nutrition improvement, cultivate research and monitoring, and mitigate and relieve the impacts of invasive species. Furthermore, under the IJC's umbrella, an array of committees, commissions, and teams have been established to screen and study shared watersheds, provide informed advice to the governments on specific issues, and participate in facilitated board meetings. Makate et al. (2018) review the water footprint methodology to increase stakeholder engagement.

Dinar (2016) focuses on the issue of water scarcity in the U.S. It mentions the gap between water quantities supply and demand resulting from the decline in available quantity and the deterioration of the quality of water resources, the need to consider water at the economy-wide level due to its key role as an inter-sectoral mechanism and the need to shift research from technological innovations to the potential of institutions in managing water resources.

Gleick (2002) calls for a new paradigm in water management. He distinguishes between the hard path, which involves cement and iron to seek sources of new supply, and the soft path, which seeks to improve the overall productivity of water use and deliver water services corresponding to the needs of end users. Gleick (2003) studied global freshwater resources and observed the soft-path solutions for the twenty-first century on "the construction of massive infrastructure in the form of dams, aqueducts, pipelines, and complex centralized treatment plants to meet human demands." Soft path solutions consider the centralized physical infrastructure that ultimately allows for decreased cost in community-scale systems, distributed and open decision-making, water markets, efficient technology, and environmental protection. One study looks at the Interbasin Water Transfer (IWT), which aims to restore an intermittent transboundary stream's ecology and improve recreational opportunities (Ayun, Israel).

Shivakoti and Bastakoti (2006) analyze the strength of Montane irrigation systems in Thailand. Araral (2009) models the incentive problems in foreign aid, focusing on how moral hazard and aid fungibility interact with incentives.

Sears et al. (2018) discuss the economics of sustainable agricultural groundwater management institutions, including the possible perverse consequences of incentive-based agricultural groundwater conservation programs. The paper highlights the importance of different types of water management/institutions: dynamic management, conjunctive management, spatial management, and property rights.

Rai et al. (2019) study on water availability in a small town in Nepal identified the ideal institutional framework for implementing an incentive payment for ecosystem services (IPES) and fund flow mechanism. Balasubramanya et al. (2018) assess the impact of training on water user association (WUA) staff on the performance of mandated duties of the WUA.

Jones-Crank (2024) conducted an institutional analysis of the water-energy-food (WEF) nexus, using interviews administered for three cities: Phoenix, Cape Town, and Singapore. The institutional analysis allows the examination of multiple levels of governance among WEF actors in the cities. The study suggests that WEF nexus governance, while present, is limited in practice. Results also suggest that each city in the study is characterized by a specific structure of WEF nexus governance.

Crespo et al. (2019) utilize a hydro-economic modeling approach to analyze the tradeoffs between economic water uses and environmental flow requirements in Spain's Ebro River basin, where rivaling regional interests have created intense political debates over water allocation. The authors examine three different water allocation policies under diverse drought scenarios, which helps evaluate how increased environmental flow demands would affect agricultural activities, urban water supplies, and regional economic benefits.

Taylor and Eberhard (2020) present a comprehensive review on the successes and failures of various institutional arrangements to protect the Great Barrier Reef. Kuhn et al. (2016), use the Lake Naivasha Basin in Kenya's Rift Valley as a hydro-economic system to assess the institutional needs of this system.

Msuya and Lalika (2018) examine the institutional challenges impeding effective integration of ecohydrology and IWRM in Tanzania's Pangani River Basin, emphasizing the complex multi-sectoral coordination. The findings demonstrate that poor inter-sectoral coordination at the field level is the primary barrier preventing watershed management integration, as different ministerial structures for water and forest management create fragmented institutional approaches. This institutional fragmentation undermines growth as highland communities depend on lowland areas for food production and employment, while lowland communities rely on highland watersheds for irrigation water. However, these interdependencies lack coordinated governance mechanisms. The authors conclude that successful watershed management integration requires addressing the diverging interests of multiple participants, like water user associations and forest conservation groups, whose competing demands for ecosystem services create institutional conflicts that compromise both economic sustainability and ecological integrity.

Al-Saidi and Hefny (2018) paper analyzes water–energy–food nexus priorities for regional cooperation in the Eastern (Blue) Nile Basin—Egypt, Sudan and Ethiopia. The basin faces challenges regarding regional cooperation and increased integration, but also opportunities due to several relative comparative advantages inherent in the uneven endowments of water, energy, and arable land resources, and the levels of economic and technological development among the three riparian states. The paper also evaluates possible regional institutional arrangements and highlights the trade-offs associated with them.

Märker et al. (2018) address the critical gap in food-energy-water nexus governance by developing two distinct institutional frameworks using the IAD framework to overcome traditional sectoral "silo-thinking" in resource management. By presenting a holistic integration framework, they discuss horizontal policy integration across sectors and a vertical integration framework building upon existing institutional structures. By analyzing German policy cases, the authors showcase that effective FEW nexus governance requires adaptive institutions capable of managing complex interdependencies between water, energy, and agricultural systems across multiple governance scales.

Wutich et al. (2012), refer to tensions between institutions and existing cultural norms of justice.

Azhoni et al. (2017a) evaluate the contextual and interconnected barriers that prevent water management institutions in Himachal Pradesh, India, from effectively adapting to climate change. The study indicates that adaptation barriers include commonly cited issues and systemic problems, like normative attitudes, trust deficits, and institutional fragmentation. The authors explain how adaptation barriers are highly interdependent and contextually determined by local socio-economic, political, and cultural factors, rather than existing as isolated challenges. Moreover, inadequate financial resources in this developing economy context are not a result of climate skepticism, as usually seen in developed countries, but from fractured resource allocation driven by electoral politics and competing agendas. By providing a systems-based approach to analyzing barriers, we can better understand adaptations to climate change to design more effective institutional interventions that actually address the root causes of adaptation failures.

Gani and Scrimgeour (2014) utilize an institutional ecological economic framework to empirically examine how governance quality affects industrial water pollution levels across OECD countries. The research demonstrates that five key governance indicators, consisting of rule of law, regulatory quality, control of corruption, government effectiveness, and accountability, are statistically significantly correlated with water pollution from industrial activities, explaining how institutional quality matters for environmental protection. The analysis reveals that better governance structures effectively reduce biochemical oxygen demand (BOD) emissions across various industrial sectors.

Thiel (2014) develops a comprehensive analytical framework to explain the scalar reorganization of water governance as institutional change, using Spain's Guadalquivir river basin as a case of decentralization dynamics. The results demonstrate how modifications in water governance are influenced by ecosystem valuations, mental models, production and transaction costs, and bargaining power within constitutionally defined action situations. The Andalusian case reveals how regional actors favored decentralization to achieve better coordination between water management and regional environmental policies, driven by changing water use patterns and technological innovations.

Tingey-Holyoak (2014) investigates how agricultural businesses respond to institutional pressures for sustainable water storage management across four different regulatory environments in Australia, applying Oliver's strategic response typology to understand resistance patterns. The results showcase how agricultural water storage practices vary significantly across institutional environments ranging from weak (South Australia) to very strong (Tasmania), highlighting disjointed policy mechanisms creating unsafe storage conditions that threaten downstream communities during floods and droughts. The authors conclude that increased manager interconnectedness and reduced environmental uncertainty predict lower resistance to water storage regulations, whereas legally coercive actions consistently produce compromised responses across all institutional environments, regardless of regulatory strength.

Pulido-Velazquez and Ward (2017) study water management institutions and approaches between the United States and the European Union, highlighting fundamental differences in policy frameworks and implementation strategies for addressing water resource challenges. By contrasting the EU's integrated Water Framework Directive, which establishes unified basin-scale management across member states, focusing on ecological status, against the fragmented U.S. approach that relies heavily on cost-benefit analysis and economic principles to guide water policy decisions. The findings demonstrate that while Europe emphasizes comprehensive ecosystem-based planning through river basin management plans and public participation, the United States has developed complex economic tools, including water markets, transferable rights, and pricing mechanisms that effectively address water scarcity and allocation challenges.

Kolinjivadi et al. (2014) introduce an institutional framework for reconceptualizing payments for ecosystem services (PES), arguing that traditional market-based PES approaches fail to account for the economic and governance issues of watershed resources. Analyzing nested institutional arrangements aligning with governance structures, the authors find that ineffective PES implementation occurs because these services hinder pure market-based transactions, forcing users to negotiate arrangements socially. Specifically, PES can help achieve IAWRM goals if used within nested governance frameworks that match institutional arrangements.

Sokile et al. (2003) analyze the institutional failures preventing effective water management in Tanzania's Rufiji Basin, where population growth and increasing water demands have produced fragmented governance with conflicts and gaps. By identifying that Tanzania's water management institutions operate in isolation, the authors explain how this separates them from effective water management. They propose establishing a comprehensive institutional framework that redistributes responsibilities from centralized state control to participatory water user associations, emphasizing engagement and adaptive governance.

Colebatch (2006) focuses on the institutional context of water governance in Australia, arguing that water management should be understood as a process of institutionalization instead of simple technical responses. By analyzing how traditional water governance structures have been challenged since the 1980s, the author explores the topics of managerial reforms, market-oriented approaches, and demands for greater public participation and accountability. This demonstrates how governing water use involves three interconnected dimensions: cognitive (what people know), normative (what they value), and regulative (how they are organized).

Villamayor-Tomas (2017) writes a chapter exploring the institutional dimensions of the water-energy nexus in Europe, specifically about Spain's irrigation sector. To shape resource allocation, water and energy institutional interactions can be analyzed to increase energy demands for agricultural water distribution and usage. Highlighting how historical policies and infrastructural investments have contributed to the current governance situation, the authors explain why irrigation communities have high autonomy in systems coordinated by river basin organizations. The relationships between all institutions involved show how water and energy governance can be managed together sustainably, especially because irrigated agriculture is energy-intensive.

Scott et al. (2011) reinforce the conventional understanding of the water-energy nexus beyond resource use relationships to study institutional challenges and policy dimensions that arise when energy and water are examined across multiple spatial scales. Through three cases from different U.S. regions, the findings demonstrate how global energy demands create localized water and environmental impacts that existing institutional frameworks struggle to address effectively. The authors elaborate how critical governance misaligns where energy resources are more transportable and suitable for local adaptation strategies than water resources; however, water availability and quality remain highly localized and disconnected from energy policy decision-making processes. The authors emphasize the need for improved coordination between water and energy policy frameworks to address global change adaptation challenges.

Khan et al. (2017) conduct an econometrics study to examine the relationship between hydroclimatic variables and economic growth by analyzing 502 country-river global basins from 1991 to 2012. The authors used both exogenous hydroclimatic factors, such as precipitation, and endogenous policy-related factors, such as institutions, to analyze their impacts on GDP growth. The findings illustrate that water-related hazards, such as droughts and floods, create significant drags on economic growth; however, the impacts disproportionately impact low-income countries and regions with high agricultural dependence.

Dadson et al. (2017) explore the multifaceted role of water-related investments in promoting economic growth through enhanced productivity while also reducing losses from water-related hazards. The authors discuss how countries with difficult hydrological conditions require significant investment in both physical water infrastructure and institutional frameworks to achieve sustainable economic growth, as inadequate water security can create poverty traps. By emphasizing that optimal economic growth requires coordinated investment across multiple sectors, the authors highlight the critical importance of integrated approaches that combine water security measures with broader institutional capacity building.

Ghimire and Griffin (2014) combine the institutional and political orientation of irrigation districts, favoring irrigation over irrigators.

Gohar and Ward (2010) evaluate how establishing formal water markets could improve agricultural productivity in Egypt's Nile River basin by better allocating scarce irrigation water resources. By analyzing scenarios where farmers could trade water rights in varying geographic areas, the results indicate that this modeled water trading could heavily boost farm incomes because water can be traded for valuable crop production uses without overusing water. The authors explain how market-based water allocation mechanisms could help address water scarcity pressures in dry agricultural regions.

Videira et al. (2009) studied the participatory modeling process in the Baixo Guadiana River Basin in Portugal, where stakeholders collaborated to develop causal maps and simulation models to address water management challenges. By utilizing a system dynamics methodology to engage diverse participant groups, the authors identify problems, pressures, and impacts affecting the river basin ecosystem. The participatory process successfully encouraged integrated perspectives on water-energy-food nexus issues, showcasing important trade-offs. For example, wastewater treatment projects reduce pollution, but they also negatively impact biodiversity by converting habitats within protected areas.

Kallis (2010) uses the coevolutionary theory to analyze how water infrastructure and consumption patterns mutually shape each other over time by using Athens, Greece as a case study for their urban development. By challenging conventional narratives that view water supply as a response to expected demand, the author discusses how new supply capacity creates feedback loops that drive higher consumption and expansion. As a result, sustainable water management requires breaking these cycles through strategic policies, such as the "soft water path" that impedes supply expansion and focuses on supplier-driven conservation.

Jogo and Hassan (2010) developed an ecological-economic model to analyze the impacts of alternative policy regimes on wetlands and human well-being in the Limpopo wetland of southern Africa. The findings display trade-offs between wetland services with policy interventions affecting both ecological integrity and economic welfare of local communities. Diversifying livelihoods creates a "double dividend" effect by simultaneously enhancing economic well-being and promoting wetland conservation, while pure conservation strategies without livelihood diversification impose high economic costs on local populations. The authors emphasize that government policies supporting livelihood diversification into non-agricultural sectors are vital for sustainable wetland management, especially as climate change predictions indicate reduced rainfall, accelerating wetland conversion pressures.

Posthumus et al. (2010) develop a framework for assessing ecosystem goods and services in English lowland floodplains, emphasizing how climate change is reshaping land management priorities from traditional agricultural production toward environmental quality. The analysis applies scenario modeling to quantify trade-offs between competing land uses in the Beckingham Marshes case study, observing six different management scenarios including agricultural production, biodiversity conservation, flood storage, and agri-environment approaches. As a result, they find counterintuitive relationships between ecosystem services, such as potential conflicts between flood storage capacity and biodiversity conservation, which highlights the dynamics between controlled flood storage and agricultural production. The methodology utilizes 14 specific indicators across 5 ecosystem functions to provide a systematic analysis of land management options, representing a new integrated approach.

Diao and Roe (2003) analyze Morocco's agricultural economy to examine how trade affects water allocation and farmer welfare, identifying that removing trade protections creates a "double-whammy" effect. This means farmers' previously protected crops suffer from lower product prices and reduced water values, forcing them to form a strong political opposition to reform. As a result, the authors propose creating tradable water user-rights as an institutional solution, allowing farmers to lease their historical water allocations to other producers at market prices.

Ruiters and Matji (2015) analyze South Africa's water infrastructure funding challenges, discovering that conventional reform approaches, like privatization, have failed to address massive investment gaps and service delivery shortfalls. The authors propose seven distinct governance models for water infrastructure financing, ranging from direct government funding to full privatization, which are designed to address different aspects of the fragmented institutional context. The study reveals that South Africa requires over R600 billion in water infrastructure investment over 20 years, but current institutional arrangements face poor coordination, inadequate cost-recovery mechanisms, and limited capacity to access financing. By proposing hybrid institutional models, the authors emphasize the need for stronger governance frameworks.

Dharmaratna (2011) examines water pricing and welfare impacts in Sri Lanka's pipe-borne water sector, analyzing demand patterns and cost structures across various sectors. The author studies under-pricing issues where water tariffs fail to recover full supply costs, resulting in overconsumption and financial sustainability issues. Using econometric modeling techniques, the authors explain how implementing marginal cost-based pricing structures could generate significant efficiency gains while balancing the economy and equity for water service provision in developing country contexts.

Jones-Crank (2024) conducted an institutional analysis of the water-energy-food (WEF) nexus, using interviews administered for three cities: Phoenix, Cape Town, and Singapore. The institutional analysis allows the examination of multiple levels of governance among WEF actors in the cities. The study suggests that WEF nexus governance, while present, is limited in practice. Results also suggest that each city in the study is characterized by a specific structure of WEF nexus governance.

3.5 Developing vs. Developed Countries

Sixteen papers describe the different types of institutional developments in both developed and developing countries. Institutional water development can vary between developed vs. developing countries. Comparing the various frameworks, interventions, institutions, and policies employed in different countries can be a step forward in improving the future of institutional water development worldwide. Grey and Sadoff (2007) study water security in both developed and developing countries and define water security as the accessibility of a satisfactory amount and nature of water for wellbeing, occupations, biological systems, and creation, combined with a worthy degree of water-related dangers to individuals, environmental conditions, and economies. According to this study, the only proven way to accomplish water security at a public/national scale has been through "investment in an evolving balance of complementary institutions and infrastructure for water management." Yet, the present water-insecure nations face more troublesome hydrologists, more extensive populations with more shifted water demands, and a greater understanding of and hence greater responsibility regarding the social and climate impacts inherent in water management. The brief examination of this investigation demonstrated no clear option in contrast to achieving water security in developing countries. Therefore, these nations should not see the water framework alone as a final solution. Without developing appropriate water institutions, poorly managed infrastructure will likely not help development; it may even prevent development. Each exertion should be committed to guaranteeing that the costly mistakes of the past are prevented in the future. Overall, Grey and Sadoff (2007) study concluded that worldwide guidance for water security is better adjusting and sequencing interests in water framework and foundations intended to adjust to changing qualities and needs, for considering every possible alternative, and afterward fitting these decisions to country-specific conditions and for pushing down the social and natural costs of accomplishing water security.

Developed countries are often expected to have more developed water institutions and policies that shape their water systems; however, this does not mean that other problems do not exist. First, we would like to focus on developing countries and the types of water use and institutions. Rusca and Schwartz (2014) study on accommodating local institutions in water governance, with a general focus on developing countries, the importance of legal pluralism, and 'going with the grain' governance approaches. Legal pluralism has to be managed by invigorating a shift from one to a confirmed connection between different frameworks' decisions that apply to a similar circumstance. This is to be accomplished by obliging or guaranteeing common help between numerous frameworks. Additionally, 'going with the grain' governance supports the fusion of socially implanted foundations into current institutional frameworks, with the understanding that this will guarantee more successful and effective improvement results.

We will discuss studies on Argentina, Nepal, Colombia, China, and Bangladesh from our selected literature bundle for developing countries. Berardo et al. (2013) introduced IWRM in Argentina in the 2000s and analyzed two main institutional initiatives primarily used to implement this water institution in the country. This study explores more of the general institutional initiatives practiced in the country and ways to improve implementation, rather than the type of water use. The two initiatives include (1) the adoption of the Guiding Principles of Water Policy and (2) the ongoing design of the National Federal Plan of Water Resources. They conduct an analysis on these new institutions through an adaptive governance framework to examine how successfully national and provincial governments have dealt with the issues of representation (who participates in decision-making processes) and process design (how choices are made) that are so important in the early phases of dealing with inter-jurisdictional water issues. Their findings confirm that creating an institutional framework to progress the integrated management of water resources in developing countries is difficult and time-consuming. Ultimately, frail bureaucracy, with little technical capability, typically coexists with scarce national resources to develop and implement critical planning efforts, and state that they "can even be used in extreme cases as the political arms of elected politicians seeking to discipline their opposition or reward allies."

In contrast, Gutiérrez-Malaxechebarría (2013) conducted research on informal irrigation in the Colombian Andes, which aimed to provide more details on the scale of informal irrigation systems, their kind, the customers they serve, and the types of water sources they use. Additionally, they created a large-scale partial inventory, questionnaires, interviews, and field trips to collect more information through statistical analysis. Their findings concluded that the "Correlation analysis indicated the following predominant characteristics of current informal irrigation systems: (1) they are carried out by independent small-scale family farmers, (2) they are individual irrigation systems that do not involve agreements between producers, (3) water is transported through hoses, (4) sprinklers are used for irrigation, and (5) there are conflicts over water because of agricultural intensification". Similarly, Li et al. (2011) study on the effectiveness of China's water abstraction policies emphasizes the significance of institutional frameworks in creating well-functioning markets. They conclude that the role of governments and institutions in establishing and maintaining a robust framework for effective market-based instruments becomes more significant when policy instruments involve administered prices or created markets.

Shah and van Koppen (2006) study on whether the institution of IWRM is fit for India asks five important questions regarding the IWRM paradigm and India that should be considered by all countries when investing in whether IWRM is an appropriate institution to use. The questions include "(1) Is water poverty in countries caused by their water scarcity? (2) Would embracing IWRM help alleviate India's water poverty? (3) Is implementing IWRM feasible in India in today's context? (4) Has implementing IWRM helped counter water scarcity and poverty in other countries with a development context comparable to India's? Furthermore, finally, (5) What should be the priorities and roadmap for improving the working of the water sector in India?" They conclude that water pricing and withdrawal permits have been challenging to implement administratively, and renaming territorial entities into river basin organizations has failed to manage river basin water. They argue that while the IWRM direct demand management model is not incorrect, it is impractical in informal water economies. The rise of a class of intermediaries between users and natural water sources, in the shape of water service providers, is a precondition for meaningful demand management. The emergence of a new class of middlemen between users and natural water sources, known as water service providers, is a prerequisite for effective demand control.

Mottaleb et al. (2019) studied water markets and irrigation services in Bangladesh, acknowledging the viability of the groundwater irrigation economy. Next, we observe two studies that focus on developed countries. Hunt (2007) overlooks communal irrigation in developed countries in a comparative perspective, stating that 1,000,000 communal irrigation systems currently exist. The study recognizes three main forms of communal management: (1) Irrigation Communities (composed of all and only the irrigators); (2) Municipalities (where the local administration does the management); and (3) Districts. Environmentally sustainable communities exist only when they are also economically viable. Finally, 'occidental despotism' is a strange occurrence in which conquering nations force state administration of water on their colonies despite having communal control at home. Satoh (2019) examines the effectiveness of water rights.

Bastakoti and Shivakoti (2012) investigate the rule system in irrigation management in Nepal. The paper analyzes the rule formation in several irrigation systems using the ADICO syntax method. The paper finds differences in rule formation and enforcement between farmer-managed irrigation systems and agency-managed irrigation systems, with stronger rule enforcement mechanisms in farmer-managed systems compared with agency-managed systems. Kumar (2018) argues that policy reforms in the water sector with the dual focus of investment in water infrastructure and in water institutions are essential for improving management of water resources in developing countries. The chapter uses recent datasets from one of the Indian states that pertain to multipurpose projects.

Bulengela (2024) questions the additional needs to achieve sustainable water management in Tanzania. The paper examines, using a systematic literature review, the experience of water resources management in Tanzania, particularly that of formal institutions. Findings indicate that formal institutions failed to ensure sustainable water resources management; they have weakened the informal institutions and exacerbated water conflicts. One of the suggestions derived from the paper is to find ways to integrate informal institutions in formal water resources management initiatives.

Mujtaba et al. (2024) present the Government of Pakistan's perspective on accomplishing sustainable water management through the UN SDG-6 roadmap and identify the associated challenges faced by the Government of Pakistan in meeting these goals. The paper identifies several possible directions, such as improving water governance, water conservation enhancement, and equitable water distribution.

Ishaque et al. (2023) argue that surface water and groundwater in Pakistan are depleting sharply, and if not addressed properly, they have the potential of becoming the biggest national security problem. The paper's analysis recommends an effective integration of smart technology in addressing multidimensional water issues and efficient water governance for ensuring water security for the sustainable development of Pakistan.

Abbas et al. (2023) discuss and compare water security and governance for the non-arid Canada and arid Qatar. Comparing the climatic and physical conditions facing these two regions allows the comparisons of water sustainability issues, dimensions, security, and governance, and support the improvement of water governance structures for resource sustainability, food security, and climate change adaptability.

Pahl-Wostl and Knieper (2023) investigate requirements for governance systems to become polycentric and to effectively tackle water management challenges, applying Qualitative Comparative Analysis (QCA) to data from 26 cases in developed and less developed countries. Results suggest that polycentric governance systems are characterized by high performance in supporting effective coordination compared to fragmented, as well as centralized, non-coordinated systems that are characterized by poor performance. In addition, the QCA analyses point to the importance of high institutional capacity for the effectiveness of formal coordination institutions.

3.6 Type of Water

Fifteen papers described the types of water use (surface water, groundwater, wastewater, conjunctive use) in institutions worldwide. Various studies have revealed that each institution follows different frameworks, parameters, and criteria in various regions.

Ananda and Aheeyar (2020) study of groundwater institutions in India selected property rights as a valuable framework for analyzing alternative institutional structures because they are often viewed as a bundle of entitlements to a resource rather than just ownership rights. Rai et al. (2019) study on water availability in a small town in Nepal identified the ideal institutional framework for implementing an incentive payment for ecosystem services (IPES) and fund flow mechanism.

On the other hand, Muchapondwa et al. (2018) revealed many lessons from applying market-based incentives in watershed management and emphasized that legal and institutional reforms are required for successful watershed cases. This study also validates the effectiveness of the IPES scheme; however, IPES is only beneficial when there are enough funds to support it. In an editorial note to a special issue on water institutions, Saleth (2018) interprets that Muchapondwa et al. (2018) study makes an interesting point from the perspective of institutional economics. The victim pay principle is not the strongest because of prevailing institutional distortions linked to the reality and usefulness of property rights.

Different types of water are behaviorally guided by their institutions. Edwards and Guilfoos (2021) study revealed that many efforts to minimize drawdown are hampered by inadequate and weak institutions. In South Asia, unregulated groundwater extraction rates have exhausted aquifers, resulting in a slew of socioeconomic, environmental, and human health issues. Rai et al. (2019) study aims to create effective, fair, and sustainable groundwater institutions regulating excessive groundwater extraction rates, which has been one of the major policy problems.

Policy recommendations and interventions play an important role in the outcomes of many institutions. However, policy recommendations depend on the region's situation, and one policy might not fit all. Hayden and Tsvetanov (2019) studied irrigation restrictions in California and found that a peak policy intervention in California was reached in April 2015 when a very dry winter left no snowpack on the slope of the Sierra Mountains. As a result of this, Governor Brown mandated a regulatory measure for a 25% reduction in residential water use across the state. In contrast, Chang et al. (2010) study on land subsidence and production efficiency in Taiwan suggests that it is important to gain a deeper understanding of the actions of producers and the factors that influence their decisions to exit the industry. Their study finds an effective policy initiative to impose extra costs on groundwater users. Cobourn (2015) study on the externalities and simultaneity in surface water-groundwater systems and challenges for water rights institutions examines how water-use decisions and improvements in irrigation technologies affect communication across space and time, posing a challenge for policy instrument design.

Kuwayama and Brozović (2013) study on the regulation of spatially heterogeneous externality and tradable groundwater permits to protect streams states their concern for groundwater pumping and how it can restrict surface water flow in surrounding streams. Stream depletion has led to litigation between local stakeholders and federal agencies in Idaho and Texas over species protected under the Endangered Species Act and species that are commercially and culturally important, such as Pacific Northwest salmon. Although groundwater use in the United States is largely unregulated and unmonitored, worries over stream depletion have resulted in water management institutions for groundwater constantly changing. This study examines whether imposing geographically varied groundwater pumping regulations leads to significant reductions in farmer abatement costs and stream damage using a population data set of irrigation wells in the Nebraska portion of the Republican River Basin. Their findings indicate that regulators can generate the majority of the possible reduction in total social costs without accounting for geographic variation. However, if regulators need to improve stream protection from existing levels dramatically, regionally diversified rules will result in significant cost savings. In addition, by adopting a one-to-one tradable permit system that does not account for geographic heterogeneity, regulators can accomplish the majority of the possible savings in total social costs (i.e., farmer abatement costs plus stream damage costs).

Golovina et al. (2023) investigate the organizational, economic, and regulatory aspects associated with groundwater extraction by individuals in the Russian Federation and compare it to the situation in Germany and China. The paper reviews the state-level legislation and regulation related to individual groundwater pumping and identifies shortcomings in the system, suggesting alternative mechanisms for legalizing the activities of individual groundwater users.

Ishaque et al. (2023) argue that surface water and groundwater in Pakistan are depleting sharply, and if not addressed properly, they have the potential of becoming the biggest national security problem. The paper's analysis recommends an effective integration of smart technology in addressing multidimensional water issues and efficient water governance for ensuring water security for the sustainable development of Pakistan.

Mirzaei and Theesfeld (2025) analyze two key groundwater laws, the Equitable Distribution of Water Resources Law and the Ta'een Taklif (Law of Iran), using Institutional Grammar, based on Attribute, Deontic, Aim, Condition. Both laws were found to be norm-based, with limited enforcement mechanisms, although with a government-centric approach, making the government the primary actor and users of groundwater passive. The paper highlights the need for a change leading to laws that maintain a balanced integration of norms and rules to create a robust and institutionally sustainable groundwater management.

3.7 International Water

Forty-nine papers described the different types of international institutional developments in water. Institutional water development can vary across different states, countries, and borders; however, there are certain commonalities, differences, and aspects of international water, such as relations, laws, conflicts, and cooperation, that all institutions pay close attention to. Frameworks, interventions, institutions, and policies all play a significant role in international water relations that can help maximize water resources, technology, and availability globally.

International water practices vary, and these theories are still being criticized today. According to Easter and McCann (2010) since there has been overinvestment in water infrastructure and underinvestment in water institutions in the past, there is a need to place a higher focus on the creation of new and better water institutions, particularly water rights.

Marshall et al. (2013) study on Australia's Murray-Darling Basin (MDB) acknowledges researchers' international use of integrated water resources management (IRWM). It expresses the need for this widely practiced theory to be re-evaluated by proving how cross-border integration challenges emerge. This study uses the MDB as a longitudinal case study to understand the cross-border integration challenges of water.

International water relations can become complicated when neighboring countries have underlying historical conflicts that impact their current cooperation efforts. Altingoz and Ali (2019) study on the Armenian-Turkish border reveals that sharing the Arpacay/Akhuryan Dam means that these countries have to maintain riparian cooperation. This cooperation between the local cross-border water institutions is made possible through polycentric management principles, which consist of "multiple centers of decision-making authority with no dominating central authority". Similarly, Syed and Choudhury (2018) study the Indus River to demonstrate the assimilation of cooperation and conflict over time.

Although some commonalities exist within approaches and frameworks to different river basins across the globe, many differences need to be addressed to understand which approach serves the overall best interest of international water relations and cooperation. Rieu-Clarke (2015) study on transboundary hydropower projects expands on the weak cooperative arrangements between neighboring countries and borders that share river basins. However, these river basins depend heavily on 'out of basin' concepts for water sharing. Under customary international law, a system of substantive and procedural laws has developed to define the rights and obligations of the states that share these transboundary rivers. Increased worldwide attention to water resource management has led to the establishment of new institutional arrangements and financing methods, and international efforts aimed at strengthening river basin institutions. Gerlak (2004) studied the Danube River and how to strengthen regional governance bodies.

Islam and Madani (2017) study on the contingent approaches to managing complex water problems highly focuses on and recommends capacity development as a critical strategy for improvement in water diplomacy. Capacity development will include efforts to enhance practitioners' professional skills in water resource management, knowledge of international water law, interpersonal communications, and conflict resolution at both the institutional and personal levels.

Dellapenna et al. (2013) study on the future of global water governance examines the relationship between global water governance structures and their responsiveness to shape projected water, specifically in the United Nations. The study suggests that for water institutions considering developing governance strategies in order to respond to emerging global water crises, policymakers must first decide what kind of water future they want, for example, how they want to ensure that the two overarching principles of global water governance, access for all and sustainable water resource management, are met. Similarly, Water (2019) study on water development in the United Nations emphasizes that community-based action is critical to address the core reasons of 'leaving people behind' in terms of water and sanitation. Water resource allocation mechanisms can be set up to meet a variety of socioeconomic policy goals, such as ensuring food and/or energy security or promoting industrial growth, but ensuring that enough water is available (and of suitable quality) to meet everyone's basic human needs (for domestic and subsistence purposes) must be a top priority. The UN member states have made commitments to adopt the 2030 Agenda for Sustainable Development and recognize the human rights to safe drinking water and sanitation, both of which are critical for reducing poverty and building successful, peaceful communities.

Although water governance has been a common institution mentioned by many selected studies, no study has exclusively focused on a water governance reform framework (WGRF). However, Grafton et al. (2019) used WGRF to establish water supply and demand convergence to maintain the sustainability of freshwater ecosystem services. The WGRF consists of seven critical strategic considerations regarding water reform and implementation of the framework, which is designed to not only be flexible for use in any country but also integrative in regard to the reform research goal, inequities in water allocation, and simple to use or flexible to many scales and contexts. The seven considerations of WGRF include: "(1) well-defined and publicly available reform objectives; (2) transparency in decision-making and public access to available data; (3) water valuation of uses and non-uses to assess trade-offs and winners and losers; (4) compensation for the marginalized or mitigation for persons who are disadvantaged by reform; (5) reform oversight and "champions"; (6) capacity to deliver; and (7) resilient decision-making that is both beneficial and durable from a broad socio-economic Perspective". The study applies WGRF to the following countries: Murray–Darling Basin (Australia), Rufiji Basin (Tanzania), Colorado Basin (USA and Mexico), and Vietnam. The study's evaluation of WGRF in these five countries proved that the framework was successful. They believe it should be the "core" of water governance reform efforts since it is adaptable enough at the local, basin, and national levels. For the Murray-Darling Basin and Australia, they provide a framework for assessing reform successes and failures, highlighting power imbalances, and, most crucially, providing constructive assistance for adaptive reform management. Second, when applied to Tanzania's Rufiji Basin, the WGRF examines the existing water governance system and, as a result, suggests future water reform options (Grafton et al. 2018).

It is also critical to address that not all water governance practices are beneficial. Thomas and Warner (2014) study on river basin multi-stakeholder platforms (MSP) and the practice of good water governance in Afghanistan examines the chosen pilots for the new models, the water management and conflict resolutions in two sub-basins in North Afghanistan, lower Kunduz, and Taloqan, and in two different dry years: 2008 and 2011. The study finds significant gaps between practices and models 7 years after implementing "good" water governance principles in Afghanistan. It is crucial to remember that putting IWRM, RBM, and MSP participation into practice with results that add value to existing processes takes time. The study also suggests that in the post-conflict context of Afghanistan, rigorous implementation of the MSP model of contemporary water management may be "counterproductive." What is gained in terms of 'good governance' compliance by a rigid application of the law may be at the expense of performance.

Middleton and Dore (2015) study on transboundary water and electricity governance in mainland Southeast Asia investigates the links and disjunctions between regional electricity and water governance frameworks and the context, drivers, instruments, and arenas of water and electricity decision making. Linkages between power administration and water administration, while for the most part frail and packed with power imbalances, are distinguished, including Ecological Effect Appraisal and Key Natural Evaluation devices. Disjunctures incorporate state sway and restricted coverage of entertainers among the power and water administration fields. In addition, Söderbaum (2015), studies transboundary water management in the Zambezi river basin.

Fernandez (2013) explores three-game theory scenarios of how transboundary water institutions are facilitated in sovereign country management of the international borders of North America through taking responsibility for pollution control expenses, sharing water monitoring responsibilities, and formal decision-making. A noncooperation game from an earlier time of unilateral decision-making, a cooperation game with water monitoring and information exchange for decisions, and a Stackelberg game with official financial channels for one nation distinct from decisions for each country's wastewater pollution reduction are the three-game theories/scenarios. Results show that cooperation reduces overall costs and damages, with the Stackelberg game having the next most significant costs and damages for the steady-state.

While there seems to be a pattern of transboundary water governance in most of our studies in the bundle of international water articles, Rak et al. (2019)studied State cooperation on implementing the ballast water management convention in the Adriatic Sea. This paper reveals, based on acquired legislative and institutional data, and taking into account regional marine traffic and environmental conditions, that via proper Adriatic States' collaboration, the integration of current environmental legislation commitments, as well as a stronger interaction between public institutions from the maritime and environmental sectors, may encourage the execution of ballast water management requirements.

A big part of international water is the treaties and laws governing and guiding institutions. A couple of studies have addressed the different areas of water concern of the Great Lakes. VanNijnatten et al. (2016) study on assessing adaptive transboundary governance capacity (TGC) in the Great Lakes Basin discusses how transboundary governance began with the 1909 Boundary Waters Treaty (BWT) and the establishment of the International Joint Commission (IJC). In reaction to unused logical data, almost genuine contamination impacts within the Bowl from industrialization and urbanization, the 1972 Great Lakes Water Quality Agreement (GLWQA) put in place additional transboundary instruments to address water contamination. Reestablished in 1978, 1987, and 2012, this "non-official, good-faith understanding between the two levels of government" comes about in a center on 43 Areas of Concern (AOC) and the execution of Remedial Action Plans to clean up and delist contaminated waters. A collaborative foundation was built up to control chemical inputs and nutrition improvement, cultivate research and monitoring, and mitigate and relieve the impacts of invasive species. Further, under the IJC's umbrella, an array of sheets, commissions, and teams have been set up to screen and study shared watersheds, to give logical advice to the administrations on specific issues, and to participate in facilitated board endeavors.

Similarly, Heinmiller (2016) study on the institutions and transboundary governance capacity in the Great Lakes Basin finds and suggests that the "Michigan Department of Environmental Quality (MDEQ) may be in a better position than the Ontario Ministry of the Environment (OME): Michigan erased its budget deficit in 2013, while Ontario was still burdened with a nearly $12 billion deficit, suggesting that OME's budget, and perhaps its governance capacity, are under fiscal threat for the foreseeable future". Gaden (2016) study on the cross-border Great Lakes fishery management examines the Joint Strategic Plan and its lake committees as a set of institutional arrangements for transboundary governance, using the four metrics described in the framework paper in this special issue: functional intensity, stability and resilience, legitimacy, and compliance. All four institutional indicators for the plan's transboundary governance capacity are high: it supports deep ongoing connections, it is robust, it is legitimate in the eyes of a strong "epistemic community" of fishery management specialists, and it has effective compliance procedures. Though integration is improving, the plan does a poor job of connecting fishery management with other Great Lakes policy goals (such as water quality improvement and habitat conservation). Greitens (2016) study on transboundary governance capacity in the Great Lakes examines AOCs' underlying transboundary governance architecture in terms of "functional intensity, nature of compliance mechanisms, stability and resilience, and legitimacy." According to the findings, transboundary governance requires more centralized enforcement and finance for the AOCs to continue to progress. The findings reveal that AOCs might have considerable transboundary governance flaws in terms of compliance procedures and ideas of stability and resilience but significant functional intensity and legitimacy. Transboundary governance in the AOCs tends to deteriorate over time, even when early governance solutions are established well with strong stakeholder engagement.

Additionally, VanNijnatten et al. (2016) discuss the aquatic invasion and assembling transboundary governance capacity for prevention and detection in the Great Lakes Basin. Similar to the previous study, they evaluate the functions and operation of institutions and networks using the following indicators: functional intensity, nature of compliance mechanisms, stability and resilience, and legitimacy. Ultimately, their results are consistent with the previous study. Transboundary architecture is functionally complex and intense, operating in the realms of collaboration and, in some cases, harmonization across both informal and formal networks. Furthermore, Friedman (2016) also uses the four indicators mentioned above and concludes that the TGC framework provides valuable insight into the diverse Arctic transboundary governance system. Although the study is in agreement with other studies that TGC is a complex system, they suggest that a network of actors operating at the same time can be the greatest "barometer of capacity". Garrick et al. (2016) utilize the TGC framework to compare responses to nonpoint contamination in the Great Lakes and transboundary water basins in North America and Australia. "Issue linkages" have been employed in the Columbia, Colorado, and Murray-Darling Basins to address governance capacity gaps by drawing on other water-related concerns, such as fisheries and water scarcity, where governance capacity exists. Issue linkages between nonpoint pollution and salmon recovery improve water quality governance, coordination, and harmonization throughout four main US states, including Oregon, Washington, Idaho, and Montana. In every case, building transboundary governance capacity has necessitated a focus on 'process values,' or the processes used to make decisions and implement nonpoint pollution initiatives. Future studies should look at how features of transboundary governance ability change over time in connection with environmental quality indicators and develop finer-grained indicators to guarantee external validity and allow comparisons between case studies.

Other studies have highlighted the importance of Water Footprint Assessment (WFA), the collective study of freshwater use, scarcity, and pollution relative to consumption, production, and trade patterns. Hoekstra et al. (2019) focus on reducing the water footprint of our production and consumption habits.

Gleick (2003) studied global freshwater resources and observed the soft-path solutions for the twenty-first century on "the construction of massive infrastructure in the form of dams, aqueducts, pipelines, and complex centralized treatment plants to meet human demands." Soft path solutions consider the centralized physical infrastructure that ultimately allows for decreased cost in community-scale systems, distributed and open decision-making, water markets, efficient technology, and environmental protection.

Addressing water conflicts and how different countries have managed to use institutions that serve their country the most. Zikos and Roggero (2013) explore the link between institutional fit and political divisions in Cyprus.

Water conflict on a broader scale becomes an international water conflict, which can become a threat to international security. One of the most significant international water conflicts considered today is climate change. Tir and Stinnett (2012) study focuses on how institutions can mitigate international water conflict. One of the direct impacts of climate change is the increase in demand for water, and the increasing number of droughts will lower the supply of water in the long term. This study notes that the propensity for conflicts over water to escalate depends on whether a formal agreement governs the river in question. Ultimately, the institutional design of river treaties will determine their capacity to respond to increased water stress caused by climate change. Tir and Stinnett (2012) study focuses on four specific institutional features: provisions for joint monitoring, conflict resolution, treaty enforcement, and the delegation of authority to intergovernmental organizations. Treaties with more institutional characteristics are likely to handle water-related disputes and conflicts better. Through analyzing historical data on "water availability and the occurrence of militarized conflict between signatories of river treaties between 1950 and 2000", they were able to test this expectation. According to the empirical findings, water shortage does raise the probability of armed conflict, but institutionalized agreements mitigate this risk. The results indirectly prove that by acting as a buffer between climate change and international security, the presence of international institutions can be an essential method of adjusting to the security repercussions of climate change.

Kuwayama and Brozović (2013) study on tradable groundwater permits to protect streams discusses how groundwater pumping can cause a reduction in surface water flow in nearby streams. This type of externality has been the cause of Intra/inter-state conflicts and a fast, constant change in water institutions and management policies. The study examines whether implementing regionally varied groundwater pumping regulations leads to significant reductions in farmer abatement costs and costs from stream damage using a population data set of irrigation wells in the Nebraska section of the Republican River Basin. Their data results demonstrate that without accounting for geographical variation, regulators can generate the majority of the possible reductions in overall societal costs.

Gleick (2003) studied global freshwater resources and observed the soft-path solutions for the twenty-first century on "the construction of massive infrastructure in the form of dams, aqueducts, pipelines, and complex centralized treatment plants to meet human demands." Soft path solutions consider the centralized physical infrastructure that ultimately allows for decreased cost in community-scale systems, distributed and open decision-making, water markets, efficient technology, and environmental protection. One study looks at the Interbasin Water Transfer (IWT), which aims to restore an intermittent transboundary stream's ecology and improve recreational opportunities (Ayun, Israel).

Al-Saidi and Hefny (2018) paper analyzes water–energy–food nexus priorities for regional cooperation in the Eastern (Blue) Nile Basin—Egypt, Sudan, and Ethiopia. The basin faces challenges regarding regional cooperation and increased integration, but also opportunities due to several relative comparative advantages inherent in the uneven endowments of water, energy, and arable land resources, and the levels of economic and technological development among the three riparian states. The paper also evaluates possible regional institutional arrangements and highlights the trade-offs associated with them.

Varady et al. (2023) traces the evolution of transboundary water scholarship and identifies five framings used in transboundary water governance and management: conflict and cooperation; hydro-politics; hydro-diplomacy; scale; and disciplinary approaches. The paper discusses possible future directions for transboundary water governance and management, identifying the need for additional research on how to deal with climate-related and other new challenges. Suleymanov (2024), reviews the institutionalization of the Kura-Aras River Basin, an international water-scarce basin in the South Caucasus.

Sehring et al. (2024) explore the politics of institutionalizing river basin management in Central Asia, the Former Soviet Republics, with a focus on the establishment of basin management organizations in Kazakhstan and Tajikistan. The authors use published academic literature, policy reports, and interviews with experts in the region. Results show that there is a close interaction between interests and activities of national and international actors, and that the influence of water governance norms that were promoted by international donors has initiated legal changes in all countries of the region, but were implemented to a lesser degree. The results of this paper overlay those of Suleymanov (2024), which refers to a different basin in the same region and finds similar reasons for failing to promote functioning institutions.

Prniyazova et al. (2025) examine the complexities of water resource governance in the Central Asia/Aral Sea region, emphasizing the relationship between national interests and regional cooperation. Ibrahim and Farah (2025), examine the international frameworks that were developed separately for climate change and for water security concerns, for both shared surface and groundwater sources.

Jibat et al. (2024) evaluate the links between institutions developed to deal with climate change and water security in international water, surface, and groundwater. The paper analyzes the perspectives of both legal regimes and international basin water agreements, and proposes ways to enhance synergies between them. The proposed enhancements are tested in the Nile River Basin and the Guarani aquifer.

Meissner et al. (2013) review the literature on water resource management institutions published between 1997 and 2011, showing that scientists are focusing predominantly on catchment management agencies and their institutionalization and organizational functionality. The paper argues that there is much less focus on other water management entities, such as advisory committees, international water management bodies, irrigation boards, the water tribunal, and water user associations.

This partial analysis of institutional management of international water management bodies leads to biased conclusions.

Turgul et al. (2025) refer to international water agreements as institutional mechanisms that can encourage cooperation over the use and management of shared water resources. They identify aspects of a treaty's design that may promote cooperation and the resolution of water-related disputes. Among these aspects they list provisions for treaty enforcement and information exchange.

Schmeier and Blumstein (2025) investigate the current state of institutionalized cooperation (via treaties and the basin organizations established by treaties) in internationally shared basins. The analysis uses a comprehensive literature review, the Transboundary Freshwater Dispute Database (TFDD) of Oregon State University (with information on international water treaties and river basin organizations). The chapter identifies various forms of conflict and cooperation over international water and the cooperation mechanisms that were developed over time, and how they were implemented and maintained. The analysis addresses the questions of whether such institutions have made a difference in sustainably and cooperatively managing, and how various future global challenges (climate change, population growth) might be addressed by these institutions.

Fowler et al. (2025) describe the evolution of the International Law Governing Transboundary Waters, with a focus on the Convention on the Protection and Use of Transboundary Watercourses and International Lakes, the UN Convention on the Law of Non-Navigational Uses of International Watercourses, and Sustainable Development Goal 6.5. The chapter analyzes three case studies that include transboundary water arrangements (a) the Orange–Senqu Region of Southern Africa, (b) the Lake Titicaca Region shared between Bolivia and Peru, and (c) transboundary water management between Canada, the United States, and Mexico. All three regions have various formal water management organizations that include dispute resolution mechanisms, but their operation varies by region and application.

Dellapenna (2025) discusses the performance of the International Joint Commission on the Great Lakes, a bi-national body with broad authority that faces difficulties in addressing its mandate due to the unwillingness of the national, subnational, and local governments to cooperate in its functioning. Therefore, this institution has seen some major successes to its credit, but also some failures. This chapter reviews the successes, which have been related to the allocation of water to competing uses across the border, and the important failures, mainly to achieve an adequate cleaning of the lakes of pollution, and out-of-Great Lakes-basin water transfers to ease water scarcity, which have been undertaken by the national governments unilaterally or by regional subnational institutions. The chapter explores the reasons for the successes and the failures. Kibaroglu (2025), reviews formal and informal water institutions in the Euphrates-Tigris Region, including Turkey, Syria, and Iraq.

da Silva et al. (2025) analyze the transboundary hydropolitical relations in the La-Plata Basin, with a focus on Brazil and Uruguay. That region has faced growing demand for water for irrigation and strong social integration. The analysis argues that the social and cultural similarities between Brazilian and Uruguayan inhabitants and rural rice producers were vital for establishing non-formal institutions to manage the transboundary waters in recent years, even when pressure on the basin's water had grown due to the expansion of irrigation. In parallel, hydrological interdependence, information sharing, and bilateral relations were key drivers in building formal transboundary institutions, in the form of the 1991 treaty signed by Brazil and Uruguay. The chapter demonstrates that cultural and historical integration among border groups is a necessary condition for effective transboundary water management institutions.

Al-Saidi (2025) introduces the various Nile Basin institutions in support of river basin development, cooperation, and conflict prevention. While these institutions have evolved in recent decades, they have been facing major challenges mainly due to failure to advance a basin-wide legal framework (treaty) for cooperation, and the increase in geopolitical disputes. The chapter proposes ways to improve the performance and stability of Nile-basin institutions, such as the use of (non-water) issue linkages.

Ashraf (2025) explores the deficiencies in the Ganges-Brahmaputra-Meghna (GBM) Basin's formal river management institutions, with a focus on India and Bangladesh's relations. The chapter argues that the existing institutions—the current international agreement (MoU) on the transboundary rivers shared by the countries, and the established bilateral river organization (such as the JRC) between India and Bangladesh on water—focus mainly on volumetric water sharing, with no attention to long-term socioenvironmental concerns, such as addressing depleting groundwater levels, increasing salinity, water pollution, and their impact on people's livelihoods.

Hjorth and Madani (2023) present a historical overview, from 1945 onwards, of events with possible impact on water management institutions, showing institutional rigidity during that period. The paper explores how an updated knowledge base could serve a quest for sustainable water governance strategies. The main message of this paper is that a persistent failure to learn is an important reason behind the dire state that the water sector is now in. The paper concludes that for a transformation of the institutional superstructure, adaptive water management (AWM) emerges as a prominent way to embark on a necessary, radical transformation of the water governance systems.

3.8   Institutions to Deal with Climate Change

Thirty-six papers described the different types of institutions that deal with climate change. Although the institutions discussed in this section vary worldwide, they all have one goal: to mitigate climate change and its effect on water availability. The progression of climate change impacts water resources, but some parts of the world are expected to experience more regular droughts, floods, wildfires, and severe heat. Water adaptation to climate change is crucial because it can affect water in many ways, such as flooding, water shortages, drinking water, water for sanitation, industry, and crop irrigation (Möllenkamp and Kastens 2012). Water governance, management, adaptation, and strategies are essential in dealing with climate change effectively. Frameworks, interventions, institutions, and policies all play a significant role in institutions dealing with climate change.

Water management and governance of institutions play a critical role in dealing with climate change impacts on this sector. With the progression of climate change, all water sources are likely to become more "variable and uncertain." Regrettably, most countries' current governance and institutional frameworks cannot address these water management and climate change-related challenges. The fundamental explanation for this institutional inertia is that most of the world's existing abstraction regimes are ineffective (Barbier 2019). With the progression of climate change, competition for limited water supplies also increases, so water regulation and governance frameworks play a critical role. Many studies have suggested that most Western governments' conventional top-down water governance system is too inflexible to deal with climate change. Mulroy (2017) study on Nebraska's water governance framework argues that there are benefits to depending on decentralized overlapping power to deal with climate change. This leads to a variety of targeted, localized solutions that foster regulatory experimentation and possibilities for jurisdictional learning.

Özerol et al. (2018) study on the comparative studies of water governance suggests that under the rising stresses of competing water uses and climate change, governance (regime) is critical to addressing water issues and transforming water management. This study then conducted a systematic review on water governance studies, in which they reflect on how water governance is defined, conceptualized, and assessed in different contexts. Their fourth area of finding relates specifically to climate change, which suggests future research to address "issues of justice, equity, and power, which are becoming increasingly important in tackling the water governance challenges that are exacerbated by the effects of climate change, industrialization, and urbanization."

Similarly, Sifundza et al. (2019) study on the El Niño drought in the Komati catchment in Southern Africa revealed exploring the different institutional and actor (30 actors-farmers, engineers, water managers, and weather services personnel) responses to the drought and the implication of these experiences for future drought management. The results of this study indicate that institutions should learn from past drought events to improve evidence-based tools, policies, and practices for drought management. How weather forecast information is packaged, communicated, deliberated, and used by institutions (e.g., KJOF) and end-users (e.g., farmers) needs to be improved to prepare for future climate change-related impacts. Their results indicate that Drought rationing is a policy measure outlined in the ISOTG guidelines that must be adjusted to the specific drought scenario. KOBWA was in charge of designing drought measures for the Maguga and Driekoppies Dams in Eswatini and South Africa, collaborating with other transboundary organizations. Lastly, Olen et al. (2016) study on irrigation decisions for major west coast crops, water scarcity, and climatic determinants conducted an econometric analysis for irrigation decisions of producers of specialty crops, wheat, and forage crops.

Transboundary water governance is not just important internationally but locally within each country's transboundary rivers, lakes, and aquifers. Transboundary water management is all about participation and cooperation over shared water assets, which is fundamental for natural sustainability, environmental change variation (climate change), local harmony and security, and economic development. One of the strategies used relatively recently by many countries is hydropower, which is a cheaper and cleaner alternative to fossil fuels and has the power to help address climate change. Rieu-Clarke (2015) study on transboundary hydropower reveals that the majority of global hydropower potential depends on transboundary rivers, but these are areas with very low cooperation. The three areas of international law include human rights, investment, and environmental protection. This study explores how these three legal regimes could be intersected to increase hydropower use in transboundary rivers. One of their main conclusions stated that foreign investors are frequently affected by disputes between states over transboundary waterways. Similarly, Ulibarri and Scott (2019) study concentrates on the National Environmental Policy Act's environmental impact assessment procedure and the Federal Energy Regulatory Commission's hydropower dam licensing process.

VanNijnatten et al. (2016) assess adaptive transboundary governance capacity in the Great Lakes basin, exploring the role of institutions and networks. Climate change is a threat to the ecosystem's health and longevity; however, it does not help that the Great Lakes are at an ecological tipping point due to being exposed to new chemicals that pose new threats and challenges to the species and even accelerate climate change. The authors highlight that those specific institutional frameworks and customary law are important for climate change, such as the United Nations Framework Convention on Climate Change and the Council for Environmental Cooperation. The Boundary Waters Treaty, the Great Lakes Water Quality Agreement, and the International Joint Commission all define the roles and duties of countries along the common boundary, as well as establish guidelines for their discussions.

Many of the papers discussed water governance and management as a critical aspect of controlling climate change and variability. The growing strain on the earth's freshwater supplies as a result of increased water use and pollution, as well as the effects of climate change, has led to widespread acceptance of the importance of freshwater in sustainable development and the essential need for improved water governance (Zhang et al. 2013). While Hoekstra (2011) study discusses why the river basin approach is no longer sufficient and argues that many of today's seemingly local water concerns have a (sub)continental or even global scope, necessitating a governance strategy that includes institutional structures beyond the river basin level. Through identifying four major issues that should be addressed at the global scale: Efficiency, equity, sustainability, and security of water supply in a globalized world, the study highlights that cooperative action is what is needed most at the global level. Possible institutional arrangements suggested are: "international protocol on full-cost water pricing and a water label for water-intensive products to the implementation of water footprint quotas and the water-neutral concept."

3.8.1 Governance and Management

Özerol (2013) conducts a multi-level analysis of irrigation management in Turkey, demonstrating how varying rules at different levels cause more environmental degradation, especially with worsening climate change. The author introduces the concepts of an "institutional scale" and "institutional alignment" to study how waterlogging and soil salinization are produced by poorly coordinated decisions, which are already exacerbated by climate change. Although participatory irrigation management was intended to promote sustainability, the lack of water rights and weak monitoring systems hinder effective farmer engagement. The author exposes how formal participation institutions can cover for imbalanced water governance that thwarts climate resilience.

Azhoni et al. (2017a) investigate climate adaptation institutions within India's water management sector, focusing on actors, inter-institutional arrangements, and barriers that slow adaptation. Through a mixed-methods approach, the authors observe how coordinated weak mandates, networks, and institutions limit adaptive action. Moreover, institutional misalignment affects their ability to plan for climate-induced water stress due to identified bureaucratic inefficiencies, poor inter-institutional coordination, and systemic governance failures. It also creates significant barriers to converting national-level adaptation policies into effective implementation.

Azhoni et al. (2017b) evaluate the contextual and interconnected barriers that prevent water management institutions in Himachal Pradesh, India, from effectively adapting to climate change. The study indicates that adaptation barriers include commonly cited issues and systemic problems, like normative attitudes, trust deficits, and institutional fragmentation. The authors explain how adaptation barriers are highly interdependent and contextually determined by local socio-economic, political, and cultural factors, rather than existing as isolated challenges. Moreover, inadequate financial resources in this developing economy context are not a result of climate skepticism, as usually seen in developed countries, but from fractured resource allocation driven by electoral politics and competing agendas. By providing a systems-based approach to analyzing barriers, we can better understand adaptations to climate change to design more effective institutional interventions that actually address the root causes of adaptation failures.

Huntjens et al. (2012) investigate institutional design principles for climate change adaptation in water management systems, building upon Elinor Ostrom's foundational work on common-pool resource management. By conducting a comparative analysis of climate adaptation strategies in three different contexts, which are the Netherlands, Western Australia, and South Africa, the authors develop eight institutional design propositions. These principles include clearly defined boundaries, proportional equivalence between benefits and costs, collective choice arrangements, and more, which are tailored for complex, large-scale water governance systems facing climate instabilities. By emphasizing the importance of "management as learning" approaches that facilitate social learning, stakeholder participation, and institutional flexibility, the authors argue that traditional institutional design principles developed for local common-pool resources are insufficient for addressing the challenges in climate change adaptation, so they developed their own.

Kirchhoff and Dilling (2016) examine the empirical application of theorized adaptive and resilient water governance characteristics across five U.S. states: Florida, Georgia, Illinois, Maryland, and Texas. The study identifies six key characteristics of effective water governance under uncertainty, which include "knowledge generation and use, policy learning with flexibility, multilevel interactions and participation, clear boundaries for water use, and incorporating nonstationary". The authors explain that successful adaptive governance means they possess these characteristics, integrating and building upon them, with transparent knowledge systems, participatory processes, and institutional arrangements combined. The authors note the gap between theoretical frameworks and practical implementation, revealing that institutional arrangements and political contexts serve as catalysts or constraints for water governance.

3.8.2 Water Market

Khan et al. (2017) conduct an econometrics study to examine the relationship between hydroclimatic variables and economic growth by analyzing 502 country-river global basins from 1991 to 2012. The authors used both exogenous hydroclimatic factors, such as precipitation, and endogenous policy-related factors, such as institutions, to analyze their impacts on GDP growth. The findings illustrate that water-related hazards, such as droughts and floods, create significant drags on economic growth; however, the impacts disproportionately impact low-income countries and regions with high agricultural dependence.

Crespo et al. (2019) utilize a hydroeconomic modeling approach to analyze the tradeoffs between economic water uses and environmental flow requirements in Spain's Ebro River basin, where rivaling regional interests have created intense political debates over water allocation. The authors examine three different water allocation policies under diverse drought scenarios, which helps evaluate how increased environmental flow demands would affect agricultural activities, urban water supplies, and regional economic benefits.

Horne and Grafton (2019) follow the development of Australia's Murray-Darling Basin water markets, tracking the evolving incremental policy transformations. Muchapondwa et al. (2018) aim to identify successful market-based incentives to help promote sustainable watershed practices.

Tir and Stinnett (2012) focus on how institutions can mitigate international water conflict. One of the direct impacts of climate change is the increase in demand for water, and the increasing number of droughts will lower the supply of water in the long term. This study notes that the propensity for conflicts over water to escalate depends on whether a formal agreement governs the river in question. Ultimately, the institutional design of river treaties will determine their capacity to respond to increased water stress caused by climate change. The study focuses on four specific institutional features: provisions for joint monitoring, conflict resolution, treaty enforcement, and the delegation of authority to intergovernmental organizations. Treaties with more institutional characteristics are likely to handle water-related disputes and conflicts better. Through analyzing historical data on "water availability and the occurrence of militarized conflict between signatories of river treaties between 1950 and 2000", they were able to test this expectation. According to the empirical findings, water shortage does raise the probability of armed conflict, but institutionalized agreements mitigate this risk. The results indirectly prove that by acting as a buffer between climate change and international security, the presence of international institutions can be an essential method of adjusting to the security repercussions of climate change.

Emmerson (2011) examines water security as an important dilemma, arguing that it is both an essential resource without substitutes and a commodity with many uses. By referencing China's recent experience with both severe drought and devastating floods as a case study, the author explains how water volatility can affect agricultural production, hydroelectric power generation, and political stability simultaneously. He also bridges the gap between traditional local water management approaches and arising global governance challenges, advocating for "glocal" solutions that combine local and global strategies.

Nüsser et al. (2019) investigates the growing vulnerability of cryosphere-fed irrigation systems relying on snow and glacier melt.

Michalak (2020) explores the effectiveness of water management policies in Polish agriculture, addressing how farms and governmental institutions adapt to climate change. He uses a dual-level analysis to assess institutional support from governmental entities and farmers' behavior through case studies, aiming to understand how Polish farms manage water resources during periods of water scarcity as climate change worsens. Despite minimal institutional assistance, larger agricultural enterprises independently adopt irrigation technologies and drought-resistant farming practices. It shows how farmers are unaware of how their water management decisions affect ecological systems and the economy.

Chen et al. (2021) evaluate water security in Xiamen, a coastal city in China that depends on freshwater resources from the Jiulong River watershed.

Eisenack (2016) analyzes institutional adaptation to water scarcity for thermoelectric power generation under climate change, using a mathematical model to compare the three regulatory arrangements based on transaction costs and environmental externalities in the German Rhine catchment. The author develops a qualitative reasoning model that analyzes how the frequency and intensity of heat waves influence the comparative costs of different institutional arrangements, finding that temperature caps perform best when heat waves increase only in intensity, while more complex arrangements become cost-effective when heat wave frequency rises. The results indicate that the introduction of Germany's minimum power plant concept following the 2003 heat wave is consistent with the model's predictions. This suggests that institutional path dependency will most likely prevent further changes despite ongoing climate change, especially since Germany is transitioning away from thermoelectric power generation.

Xie and Zilberman (2016) developed a formal analytical model to examine how institutional rules, environmental variability, and operational conditions influence decisions about the sizing of irrigation water infrastructure.

Garrido (2007) examines the performance of water markets in climatic instability through experimental economics methodology, testing how different market institutions respond to extreme wet-dry cycles. Using laboratory experiments that simulate trading periods, as modeled after Spain's Guadalquivir River Basin, the author focuses on the effectiveness of water storage mechanisms and trading restrictions during periods of severe supply variability. The findings exemplify that allowing users to carry water rights over time significantly reduces price volatility and reservoir level fluctuations during critical drought-flood cycles, whereas market restrictions intended to protect urban water users actually generate losses without protection. Through systematic analysis of market behavior during extreme climatic events rather than average conditions, the results indicate that less valuable water users play a crucial buffer by increasing consumption during abundant periods and reducing usage during scarcity.

Akron et al. (2017) conducted an economic analysis of an interbasin water transfer (IWT) project designed to restore the Ayun stream in Israel, which had dried up during dry seasons due to upstream water diversions in Lebanon. Using a time-series regression analysis and travel cost methodology, they evaluate the project's impact on recreational visitation and its associated economic benefits, finding that an increase in water flow correlated with an increase in monthly visitors. As the article provides empirical evidence that strategic water reallocation can serve as an effective management tool, we can better maintain ecosystem services in water-stressed Mediterranean regions where climate change is expected to intensify seasonal variability.

Kahil et al. (2015) evaluate the water market and irrigation subsidy policy approaches to encourage irrigation adaptation to climate change in Southern Europe. As climate change grows, it will negatively affect irrigation and water-dependent ecosystems, but the extent of these impacts depends on government policy and farmers' adaptive strategies. The authors conclude that market-based trading tends to generate better economic outcomes compared to technological subsidies through the promotion of efficient water distribution. However, they also warn of tradeoffs where markets could reduce ecological water availability, negatively impacting river flows and groundwater reserves.

Posthumus et al. (2010) develop a framework for assessing ecosystem goods and services in English lowland floodplains, emphasizing how climate change is reshaping land management priorities from traditional agricultural production toward environmental quality. The analysis applies scenario modeling to quantify trade-offs between competing land uses in the Beckingham Marshes case study, observing six different management scenarios including agricultural production, biodiversity conservation, flood storage, and agri-environment approaches. As a result, they find counterintuitive relationships between ecosystem services, such as potential conflicts between flood storage capacity and biodiversity conservation, which highlights the dynamics between controlled flood storage and agricultural production. The methodology utilizes fourteen specific indicators across five ecosystem functions to provide a systematic analysis of land management options, representing a new integrated approach.

Wyborn et al. (2023) investigate the politics of adaptive water governance in Australia's Murray-Darling Basin. Martínez-Valderrama et al. (2023) study the increasing "water gap" between agricultural water demand and available water resources, particularly in dryland regions that experience intensified water stress due to climate change. The authors argue that traditional supply-side solutions have been inadequate and even counterproductive, creating cycles of increased demand rather than addressing the underlying consumption behaviors. By using causal diagrams from System Dynamics to analyze the complex feedback mechanisms that perpetuate water scarcity, we can understand how supply-side solutions can paradoxically worsen the problem, which is coined as the "reservoir effect." Instead, they present comprehensive demand-side management strategies that discuss local adaptive practices to global policy reforms, highlighting that climate change adaptation requires a multi-pronged approach.

Zinabu et al. (2024) analyze the institutional capacity and water quality modeling challenges in Ethiopia's Awash Basin, focusing on the gap between available data and the practical implementation of in-stream water quality models for pollution control. The authors assessed seven governmental institutions responsible for water quality management, indicating significant deficiencies in database infrastructure, technical expertise, and modeling capabilities that interfere with effective water resource protection. They also evaluate six water quality models using locally relevant criteria, ultimately recommending QUAL2KW and INCA as the best suited to the basin's conditions.

Ibrahim and Farah (2025), examine the international frameworks that were developed separately for climate change and for water security concerns, for both shared surface and groundwater sources. Gharib et al. (2024) reviews formal and informal water institutions in the Euphrates-Tigris Region, including Turkey, Syria, and Iraq.

3.9   Means of Institutional Intervention

One hundred and nine papers described the different types of means of institutional intervention. This section is the longest section of this paper, which expresses what different water institutions worldwide have done to intervene and solve water conflicts. One indicator of a successful water institution is the means of institutional intervention and the ways institutions intervene during times of conflict to achieve optimal results. Institutional interventions are formalized organizational agreements between the project's management players. They are changes to the current institutional order because they alter or redraw organizational boundaries (Verweij et al. 2014). There are so many different types of intervention strategies and methods that have been used and are specific to different sectoral institutions (irrigation, residential, industrial, environmental). Not all intervention methods share the same purpose and are not necessarily the best fit for all. The most common themes present in this bundle of articles include the role of institutions and economics, IWRM, environmental cooperation, global freshwater resources, soft path solutions, irrigation development and technology, water governance, and water use associations.

3.9.1 Legal (Property and Water Rights, Water Quotas, Standards)

Madani and Dinar (2013) analyze exogenous regulatory institutions for sustainable CPR management through a groundwater case study, comparing and contrasting four external intervention types: quota-based management, resource status-based management, tax-based management, and bankruptcy management institutions. By using a three-farmer numerical groundwater example over a 50-year planning horizon, the authors can understand their effectiveness across different user behavioral characteristics based on three criteria, which are social welfare maximization, social justice enforcement, and institutional strength. The findings reveal that quota-based and status-based management institutions consistently outperform tax-based approaches in preventing "tragedy of the commons" scenarios, promoting sustainability and equity.

Jyotishi and Rout (2005) investigate water rights and resource management in India's Deccan region by analyzing the Baliraja movement and other institutions. The Baliraja movement maintained an idealistic approach of equitable water distribution, and they believed in equal water rights for all households regardless of the landholding size. By comparing the Baliraja movement and the more practical and cooperative water institutions that thrived, the results indicate the movement's failure. While it aimed to address equity, efficiency, and sustainability through restricted crop choices and equal water allocation, the movement ultimately failed because it limited farmers' economic opportunities and crop selection, especially with sugarcane. In contrast, the other successful water institutions prioritized water supply for high-value crops like sugarcane, creating practical arrangements despite inequalities. The authors conclude that policymakers must choose between equity-sustainability or equity-efficiency solutions.

Emerick and Lueck (2015) analyze the structure of water transactions in California by analyzing the determinants of contract duration using data from 1987 to 2008, with a focus on three contract types: short-term leases, long-term leases, and permanent ownership transfers. By applying transaction cost economics theory to test predictions about water market organization, the authors find that long-term agreements occur for asset-specific investments, that buyers face uncertain water supplies despite sellers having more rights, but limited evidence to show that third-party transfer impacts are minimal.

Lefebvre et al. (2012) explore whether security-differentiated water rights or single security systems improve water market performance. The authors use an experimental design with different transaction cost arrangements for a water rights market versus a water allocation market, testing how security differentiation affects allocative efficiency and risk management. When transaction costs are higher in the water allocation market than the water rights market, the findings demonstrate that different water rights increase overall profits and improve market efficiency, suggesting that the cost structure between markets is crucial.

Diao and Roe (2003) analyze Morocco's agricultural economy to examine how trade affects water allocation and farmer welfare, identifying that removing trade protections creates a "double-whammy" effect. This means farmers' previously protected crops suffer from lower product prices and reduced water values, forcing them to form a strong political opposition to reform. As a result, the authors propose creating tradable water user-rights as an institutional solution, allowing farmers to lease their historical water allocations to other producers at market prices.

Easter and McCann (2010) explain how there has been overinvestment in water infrastructure and underinvestment in water institutions historically. They alternatively propose the need to place a higher focus on the creation of new and better water institutions, particularly with water rights and their management.

Santos et al. (2023) highlight various factors that influence water management in Portugal and Brazil. The factors include robust legal frameworks, socio-economic disparities, cultural practices, agricultural water usage, knowledge sharing, public participation, climate change resilience, water scarcity risks, industrial water consumption, and urbanization. The authors conducted a SWOT analysis of water management strategies, analyzing the legal frameworks, policies, and implemented strategies in both countries.

Mirzaei and Theesfeld (2025) analyze two key groundwater laws, the Equitable Distribution of Water Resources Law and the Ta′een Taklif Law of Iran. Both laws were found to be norm-based, with limited enforcement mechanisms, although with a government-centric approach, making the government the primary actor and users of groundwater passive. The paper highlights the need for a change leading to laws that maintain a balanced integration of norms and rules to create a robust and institutionally sustainable groundwater management.

Jibat et al. (2024) evaluate the links between institutions developed to deal with climate change and water security in international water, surface, and groundwater. The paper analyzes the perspectives of both legal regimes and international basin water agreements, and proposes ways to enhance synergies between them. The proposed enhancements are tested in the Nile River Basin and the Guarani aquifer.

Al-Saidi (2025) introduces the various Nile Basin institutions in support of river basin development, cooperation, and conflict prevention. While these institutions have evolved in recent decades, they have been facing major challenges mainly due to failure to advance a basin-wide legal framework (treaty) for cooperation, and the increase in geopolitical disputes. The chapter proposes ways to improve the performance and stability of Nile-basin institutions, such as the use of (non-water) issue linkages.

Ménard (2017) reviews the variety of arrangements, coined as 'meso-institutions. Li et al. (2017) analyze how farmers' decisions on irrigated land are influenced by uncertainty in water supply, focusing on the priority of water rights systems. Hoekstra (2011) expresses how the River Basin Approach is not always sufficient. Camkin and Neto (2016) use a set of essential water issues to analyze the rights and duties of diverse players in water governance

Garrick and Aylward (2012) examine how transaction costs impact the institutional performance of market-based environmental water allocation in the Columbia Basin, which historically has over-allocated for agricultural use. Yang et al. (2003) examine China's experimental reforms in water rights and markets, specifically in the irrigation sector as well.

Ananda and Aheeyar (2020) apply a water rights framework analysis to compare the performance of all groundwater aquifers in India. Satoh (2019) examines the effectiveness of water rights. Cobourn (2015) analyzes the externalities and simultaneity in surface water-groundwater systems and challenges for water rights institutions, examines how water-use decisions and improvements in irrigation technologies affect communication across space and time, posing a challenge for policy instrument design. Brent (2017) analyzes the value associated with the system of prior appropriation in the Western United States by estimating the demand for security in water rights in the Yakima River Basin in Washington State. Garrick et al. (2011) analyze the evolution of institutional frameworks governing environmental water allocation in the Western United States.

3.9.2 Policy (Pricing, Subsidies [Direct and Indirect], Taxes, Trade, Market-Based Incentives)

Mulroy (2017) investigates climate change's impact on water management across the United States through case studies that focus on the inadequacy of traditional models in addressing contemporary water challenges. The author analyzes multiple institutional landscapes where various governance approaches are being tested, ranging from Nebraska's unique decentralized Natural Resources Districts (NRDs) system that utilizes reasonable use/correlative rights principles instead of prior appropriation, to large-scale public-private partnerships as seen in San Diego's desalination project. By focusing on the fundamental tension between environmental regulations and water infrastructure operations, as seen through the Sacramento-San Joaquin River Delta case, the author shows how restrictive regulatory approaches have failed to aid endangered species. The institutional interventions proposed emphasize collaborative adaptive management strategies, improved groundwater quantification, and the transition from traditional legal doctrines toward more flexible governance frameworks as climate change worsens.

Meinzen-Dick (2007) examines the experience of searching for panaceas in the irrigation sector. Layzer and Schulman (2013) evaluate the adoption of the IWRM in the U.S., using the Chesapeake Bay Program (CBP) as an example. Kuhn et al. (2016) use the Lake Naivasha Basin in Kenya's Rift Valley as a hydro-economic system to assess the institutional needs of this system. Horne and Grafton (2019) follow the development of Australia's Murray-Darling Basin water markets, tracking the evolving incremental policy transformations. Binz et al. (2016) investigate the institutional processes of the legitimization of reusable potable water in California. Ghimire and Griffin (2014) combine the institutional and political orientation of irrigation districts, favoring irrigation over irrigators.

Nauges and Whittington (2019) evaluate social norms information treatments (SNITs) in municipal water supply systems, addressing welfare implications. The authors develop and present a four-piece framework that considers implementation costs, utility cost savings, household welfare effects, and environmental benefits, revealing that SNITS may widely fail social benefit-cost tests because of modest treatment effects and potential costs imposed on households. While SNITs and price increases achieve similar short-term water conservation, price increases generate significantly higher net social benefits, especially in low and middle-income countries where water is usually priced below cost. The authors challenge SNITs by displaying that they should only be used as temporary drought management tools instead of permanent conservation policies, arguing that it takes away from politically challenging tariff reforms.

Kremer et al. (2011) evaluate spring protection interventions in rural Kenya, focusing on the health and economic implications of water infrastructure. The authors find that spring protection "reduces source water fecal contamination by 66%" and "decreases child diarrhea by 25%". Furthermore, households' willingness to pay for these improvements was significantly lower than standard health predictions, suggesting highly income-elastic demand for preventive health measures. The results also explain that current communal property norms, although limiting private investment incentives, may actually increase social welfare more than privatized regimes because of the inefficiencies of above-marginal-cost pricing.

Bolognesi and Pflieger (2019) develop a typology of coherence within institutional resource regimes (IRRs) and then apply it to Swiss water supply management, addressing the relationship between governance scope and coherence in policy. The authors propose eight types of coherence, such as relevance to public policies or property rights, extent-related versus discrete, and operating horizontally versus vertically, emphasizing how transversal transaction costs (TTCs) are because of institutional complexities as regions expand. The findings demonstrate that coherence between public policies, property rights, and extent-related goals causes significant institutional friction, challenging conventional assumptions about the reasons for institutional incoherence in polycentric governance systems.

Gleick (2003) studied global freshwater resources and observed the soft-path solutions for the twenty-first century on "the construction of massive infrastructure in the form of dams, aqueducts, pipelines, and complex centralized treatment plants to meet human demands." Soft path solutions consider the centralized physical infrastructure that ultimately allows for decreased cost in community-scale systems, distributed and open decision-making, water markets, efficient technology, and environmental protection. One study looks at the Interbasin Water Transfer (IWT), which aims to restore an intermittent transboundary stream's ecology and improve recreational opportunities (Ayun, Israel).

Rogers et al. (2015) provide an institutional analysis of water innovations in Melbourne, researching how various institutional works facilitate the development of institutional innovations. The three different institutional works they examine are cultural-cognitive, normative, and regulative. By looking at case studies of desalination, wastewater recycling, and stormwater harvesting, the authors explain how successful innovations only occur with institutional alignment in established governments. Citing Lawrence and Suddaby's institutional work concept and transitions theory, the authors argue how an agency-centric lens creates the institutional foundations necessary for technological change in infrastructure to improve sustainability goals.

Kolinjivadi et al. (2014) introduce an institutional framework for reconceptualizing payments for ecosystem services (PES), arguing that traditional market-based PES approaches fail to account for the economic and governance issues of watershed resources. Analyzing nested institutional arrangements aligning with governance structures, the authors find that ineffective PES implementation occurs because these services hinder pure market-based transactions, forcing users to negotiate arrangements socially. Specifically, PES can help achieve IAWRM goals if used within nested governance frameworks that match institutional arrangements.

Smajgl et al. (2009) use agent-based modeling to understand the institutional impacts of introducing water trading schemes in outback Australia. The authors examine how different arrangements for water extraction restrictions affect both environmental and economic outcomes, showing how restrictions geared toward newcomers result in substantially lower total groundwater extraction. This exemplifies the concept of institutional ripple effects, demonstrating how formal market-based water management systems can undermine existing informal community-based governance that maintained sustainable extraction.

Regnacq et al. (2016) study how physical distance and institutional barriers create friction in California's water markets, finding that water trading is geographically predictable because most exchanges occur locally due to transactional costs over distance. The analysis demonstrates that both variable costs, like transportation, and fixed costs, like regulatory approval processes, significantly limit market participation and trading. The authors conclude that policy reforms focused on reducing bureaucratic barriers and improving market information systems could improve water allocation efficiency, especially during scarce periods.

Zuo et al. (2016) develop innovative methods to measure price elasticities for water entitlements in Australia's Murray-Darling Basin, the most developed water market system. They combine preference survey data with preference market transaction data, which overcomes the traditional restrictions of data and endogeneity issues in elasticity studies. By surveying irrigators using a contingent behavior approach, the authors find that both "demand and supply of high security water entitlements are relatively inelastic within the relevant market price range of AUD $1700–$2100 per megaliter, with supply being more inelastic than demand". The results demonstrate that irrigators are generally unresponsive to price changes in their water trading decisions and tend to overestimate their willingness to sell water. As a result, the authors argue that limited price responsiveness suggests that traditional market-based approaches may struggle to reach water reallocation goals, necessitating alternative policy solutions.

Gohar and Ward (2010) evaluate how establishing formal water markets could improve agricultural productivity in Egypt's Nile River basin by better allocating of scarce irrigation water resources. By analyzing scenarios where farmers could trade water rights in varying geographic areas, the results indicate that this modeled water trading could heavily boost farm incomes because water can be traded for valuable crop production uses without overusing water. The authors explain how market-based water allocation mechanisms could help address water scarcity pressures in dry agricultural regions.

Kallis (2010) uses the coevolutionary theory to analyze how water infrastructure and consumption patterns mutually shape each other over time by using Athens, Greece as a case study for their urban development. By challenging conventional narratives that view water supply as a response to expected demand, the author discusses how new supply capacity creates feedback loops that drive higher consumption and expansion. As a result, sustainable water management requires breaking these cycles through strategic policies, such as the "soft water path" that impedes supply expansion and focuses on supplier-driven conservation.

Chang et al. (2010) study land subsidence and production efficiency in Taiwan, suggesting that it is important to gain a deeper understanding of the actions of producers and the factors that influence their decisions to exit the industry. Their study finds an effective policy initiative to impose extra costs on groundwater users since policy interventions play an important role in the outcomes of many institutions.

Thomas (1995) investigate how regulatory frameworks target pollution control despite information being asymmetrical between regulators and polluters regarding waste generation. Results find that the imperfect pollution tax or other similar policies did not effectively impact pollution levels, so they suggest implementing a "contract-based regulatory scheme". The authors also emphasize how the pollution standards and effectiveness depend on the industries, like chemical industries, which are more efficient than food manufacturers.

Ruiters and Matji (2015) evaluate South Africa's water infrastructure financing challenges by proposing seven governance models to address the country's R600 billion water investment deficit, which also improves access to basic water services. By identifying institutional barriers that prevent adequate water infrastructure development, the authors develop these governance models that range from direct government fiscal allocation (Model 1) to full private development arrangements (Model 7), including other options like water user-rights markets. The study advocates for a hybrid approach that consolidates national water resources management with infrastructure by strengthening water boards.

Zhang and Oki (2023) study China's nationwide agricultural water pricing reform for sustainable water resource management in developing countries. The research reveals that China has implemented an approach that combined "reasonable pricing" mechanisms with "precise subsidies and water-saving incentives," including investing in infrastructure and quota control management systems. China uniquely focused on farmers' affordability, targeting operation and maintenance (O&M) cost recovery to eventually progress toward full water supply cost pricing. The results display substantial water conservation achievements, even with varying progress across China's provinces, emphasizing regional disparities in economic development and infrastructure capacity. The authors explain how China balances food security concerns with sustainable water management by focusing on comprehensive policy integration instead of just relying on price adjustments to achieve conservation.

Seijger and Hellegers (2023) introduce the concept of "reorientation" to analyze how societies reform their agricultural water management systems in response to changing societal priorities about water, agriculture, and environmental concerns. The authors define reorientation as "a shift in broader societal priorities that drives reform of agricultural water management," setting this phenomenon as a social process that occurs over decades to bridge the gap between agricultural change and strategy-specific optimization. By examining 21 global reorientation examples, the authors indicate the diversity in starting points of priorities, the extent of successful implementation, and the role of government intervention. The authors identify that reorientations toward agricultural expansion and intensification appear to be more easily accommodated than those for environmental conservation, suggesting that social and biophysical boundaries limit agricultural water management reformations.

Bulengela (2024) questions the additional needs to achieve sustainable water management in Tanzania. The paper examines, using a systematic literature review, the experience of water resources management in Tanzania, particularly that of formal institutions. Findings indicate that formal institutions failed to ensure sustainable water resources management; they have weakened the informal institutions and exacerbated water conflicts. One of the suggestions derived from the paper is to find ways to integrate the informal institutions in formal water resources management initiatives.

Mujtaba et al. (2024) present the Government of Pakistan's perspective on accomplishing sustainable water management through the UN SDG-6 roadmap and identify the associated challenges faced by the Government of Pakistan in meeting these goals. The paper identifies several possible directions, such as improving water governance, water conservation enhancement, and equitable water distribution.

Golovina et al. (2023) investigate the organizational, economic, and regulatory aspects associated with groundwater extraction by individuals in the Russian Federation and compare it to the situation in Germany and China. The paper reviews the state-level legislation and regulation related to individual groundwater pumping, and identifies shortcomings in the system, suggesting alternative mechanisms for legalizing the activities of individual groundwater users.

Bilalova et al. (2023) question whether Integrated Water Resources Management (IWRM) implementation actually improves water sustainability outcomes, analyzing correlations between IWRM scores and various indicators across 124 countries. The results show that IWRM generally has positive associations with most water sustainability measures, including wastewater treatment and water-use efficiency, but surprisingly also correlates positively with water stress levels. However, the analysis reveals that contextual factors like governance quality, economic development, and environmental conditions strongly influence water outcomes more than IWRM implementation.

Ishaque et al. (2023) argue that surface water and groundwater in Pakistan are depleting sharply, and if not addressed properly, have the potential of becoming the biggest national security problem. The paper's analysis recommends an effective integration of smart technology in addressing multidimensional water issues and efficient water governance for ensuring water security for the sustainable development of Pakistan.

Martín Velasco et al. (2023) apply the OECD Water Governance Indicator Framework to the General Pueyrredón Municipality (GPM). Sowby and South (2023) compare two utilities whose novel pricing structures reflect water availability and reasonable use expectations.

Agarwal et al. (2023) analyze Singapore's water pricing reforms using household consumption data, evaluating the effectiveness of market-based conservation policies. The results display that announcing a 30% price increase generated larger conservation effects than the actual implementation, in which public housing residents reduced consumption by 3.7% more than private apartment residents due to the policy announcement alone. While higher-consumption households reacted based on the actual price implementation, households with lower usage responded more strongly to the announcements, suggesting that information salience impacts the effectiveness of policies.

Martínez-Valderrama et al. (2023) explore the widening gap between agricultural water demand and renewable water supply, demonstrating how traditional supply-side solutions paradoxically exacerbate water scarcity issues they are intended to solve. The authors explain how conventional approaches create self-reinforcing cycles where increased water availability generates greater demand, ultimately leading to resource depletion and ecosystem collapse, as seen in the Aral Sea case study. The findings deconstruct four vital mechanisms driving water gap expansion, which include the "reservoir effect" that discourages demand management, the Jevons Paradox in irrigation efficiency reforms, the growing competition from non-food crops used for biofuels and livestock feed, and the prioritization of mass production over sustainability. The authors conclude that closing the agricultural water gap requires a multi-pronged policy approach tailored to both local and global environments, aiming to move toward comprehensive demand management and sustainable resources.

Ouyang et al. (2024) evaluate China's water resources tax policy pilot program implemented across ten provinces from 2016 to 2019, defining a notable shift from the traditional fee-based system to a stringent tax-based approach for water governance and conservation. By combining theoretical economic analysis and field investigations, the authors assess the effectiveness of the Tax-for-Fee reform to demonstrate how water resources taxation influences consumption patterns through price mechanisms and behavioral changes toward alternative water sources, such as recycled water. The findings identify success factors for water tax implementation, such as stakeholder interest balance and flexible local government discretionary powers.

Akron et al. (2017) conducted an economic analysis of an interbasin water transfer (IWT) project designed to restore the Ayun stream in Israel, which had dried up during dry seasons due to upstream water diversions in Lebanon. Using a time-series regression analysis and travel cost methodology, they evaluate the project's impact on recreational visitation and its associated economic benefits, finding that an increase in water flow correlated with an increase in monthly visitors. As the article provides empirical evidence that strategic water reallocation can serve as an effective management tool, we can better maintain ecosystem services in water-stressed Mediterranean regions where climate change is expected to intensify seasonal variability.

Van den Hurk et al. (2014) compare flood risk management approaches between the Netherlands and the United Kingdom (UK), specifically examining hospital cases in these flood-prone areas. Using the IAD framework, the authors find that the Netherlands gradually incorporated several elements of the British risk-based approach. While the UK traditionally emphasizes spatial planning, the Netherlands historically focused on reducing the risk of flooding through high safety standards, demonstrating that the Netherlands is moving toward an integrated approach instead.

Saleth et al. (2016)examine the global irrigation sector by focusing on the institutional and infrastructural mechanisms required to manage growing agricultural water scarcity. Araral (2009) models the incentive problems in foreign aid, focusing on how moral hazard and aid fungibility interact with incentives. Mogomotsi et al. (2018) explore Botswana's water supply institutions and allocations, showing how their water scarcity results from physical barriers and outdated institutional frameworks for sustainable management. Shah and Narain (2019) dispute the GOI's continued arguments of water crisis, while ignoring the causes of the water crisis.

Punjabi and Johnson (2019) dissect the complex institutional dynamics of rural and urban water conflicts in India to understand how water governance shapes resource allocation between sectors. By comparing Mumbai and Chennai, they observe how Mumbai uses a historical framework of prior appropriation, allowing urban authorities to control rural water resources, which disadvantages tribal irrigation communities. However, Chennai allocates water depending on the market and riparian rights-based institutions, leading to the commercialization of groundwater, which means the usage of unsustainable extraction practices.

Li and Zhao (2018) investigate how utilizing water-efficient irrigation technologies can unintentionally cause increased water use. Dridi and Khanna (2005) analyze how asymmetric information between farmers and regulators influences water allocation, adoption of irrigation technology, and water trading.

Thomas and Warner (2014) examine river basin multi-stakeholder platforms (MSP) and the practice of good water governance in Afghanistan, examining the chosen pilots for the new models, the water management and conflict resolutions in two sub-basins in Northern Afghanistan, lower Kunduz, and Taloqan, and in two different dry years: 2008 and 2011. The study finds significant gaps between practices and models 7 years after implementing "good" water governance principles in Afghanistan. It is crucial to remember that putting IWRM, RBM, and MSP participation into practice with results that add value to existing processes takes time. The study also suggests that in the post-conflict context of Afghanistan, rigorous implementation of the MSP model of contemporary water management may be "counterproductive." What is gained in terms of 'good governance' compliance by a rigid application of the law may be at the expense of performance.

Hoekstra et al. (2019) focus on reducing the water footprint of our production and consumption habits. Bakker et al. (2008) analyze Jakarta's urban water supply systems by understanding failures in governance, highlighting how both institutional structures and ownership provide access to water for poor households. Gerlak (2017) emphasizes the importance of participation and engagement of stakeholders in river basin institutions. Makate et al. (2018) review the water footprint methodology to increase stakeholder engagement.

Rak et al. (2019) explore state cooperation on implementing the ballast water management convention in the Adriatic Sea. This paper reveals, based on acquired legislative and institutional data, and taking into account regional marine traffic and environmental conditions, that via proper Adriatic States' collaboration, the integration of current environmental legislation commitments, as well as a stronger interaction between public institutions from the maritime and environmental sectors, may encourage the execution of ballast water management requirements.

Peat et al. (2017) examine how water management agencies can create institutional flexibility to effectively implement adaptive management for aquatic ecosystem restoration. Wan et al. (2018) explore why China has failed to develop comprehensive ballast water management policies to prevent invasive aquatic species. Kajisa and Dong (2017) studied the Indus River to demonstrate the assimilation of cooperation and conflict over time. Söderbaum (2015) studies transboundary water management in the Zambezi river basin.

Li et al. (2011) investigate the effectiveness of China's water abstraction policies, emphasizing the significance of institutional frameworks in creating well-functioning markets. They conclude that the role of governments and institutions in establishing and maintaining a robust framework for effective market-based instruments becomes more significant when policy instruments involve administered prices or created markets.

Makate et al. (2018) review the water footprint methodology to increase stakeholder engagement. Tiger et al. (2011) studied the implications of residential irrigation metering on water demand and consumer expenditures in North Carolina. Cody (2019) explores how internalized cultural norms influence performance and the governance of irrigation systems in the Upper Rio Grande Basin. Hayden and Tsvetanov (2019) investigate irrigation restrictions in California and the effectiveness of the outdoor irrigation water restrictions. Behera and Mishra (2018) evaluate the performance of the long-known water management institution in Panchayat, India. Layzer and Schulman (2013) evaluate the adoption of the IWRM in the U.S., using the Chesapeake Bay Program (CBP) as an example. Tortajada and Joshi (2013) describe the development, by the city-state of Singapore, of a comprehensive plan for the overall management of its water resources. Muchapondwa et al. (2018) aim to identify successful market-based incentives to help promote sustainable watershed practices. Taylor and Eberhard (2020) present a comprehensive review of the successes and failures of various institutional arrangements to protect the Great Barrier Reef.

Söderberg (2016) analyzes Swedish water governance as a case to enhance the understanding of policy implementation in complex governance structures: how does policy coherence (or the lack thereof) affect policy implementation in complex governance structures? Such question is the result of evidence that many EU member states, including Sweden, have not reached the ecological goals for water in 2015, which also holds for the EU Water Framework Directive (WFD) system of environmental quality standards (EQS), which are part of a complex multi-level institutional landscape, embracing both EU, national, and sub-national levels. The paper maps out the formal structure of the water governance system of Sweden, focusing on power directions within the system and analyzing policy coherence.

3.9.3 *Organizational (User Organization, Decentralization, Enhancing Capacity)*

Lubell and Edelenbos (2013) examine Integrated Water Resources Management (IWRM) implementation across 13 countries to understand how different governance structures facilitate or deter water policy integration. The authors identify three distinct dimensions of integration: functional, societal, and institutional, arguing that these forms create inherent trade-offs instead of complementary outcomes. While countries with centralized systems demonstrate superior functional integration through top-down coordination, those with decentralized governance structures achieve better societal integration through increased participation, suggesting that the best version of IWRM requires hybrid governance of centralized authority and decentralized collaboration. The authors conclude that IWRM effectiveness depends on the intersection of a country's stage of economic development and the extent of administrative decentralization, arguing for an "adaptive walk" approach that recognizes fragmentation as something that continuously must be managed.

Marshall et al. (2013) conduct a longitudinal case study to understand nearly a century of cross-border water integration efforts in Australia's Murray-Darling Basin, reframing Integrated Water Resources Management (IWRM) as a collective action problem within polycentric governance systems. The authors identify three distinct phases of institutional intervention, which are the River Murray Waters Agreement (1914–1992) that focused on engineering works and water sharing, the Murray-Darling Basin Initiative (1992–2007) that emphasized collaborative inter-governmental consensus, and the Commonwealth Water Act of 2007 that established centralized federal authority. The findings demonstrate that coordination mechanisms have become increasingly ineffective as water management has grown more complex, allowing the authors to argue that successful integration now depends more on governance arrangements. Moreover, polycentric water governance systems need to balance between centralized intervention and bottom-up coordination, suggesting that governments must transition from direct control to creating institutional conditions for productive relationships and collective action.

Altingoz and Ali (2019) examine how Armenia and Turkey maintain cooperative water management through the shared Arpacay/Akhuryan Dam despite severe international conflicts, closed borders, and the absence of diplomatic relations. By using the polycentric governance theory, the authors analyze how local cross-border water institutions maintain operational cooperation through the Permanent Water Commission, which allocates water resources equally between the two nations based on agreements dating back to the Soviet era. While polycentric management enables technical cooperation at the local level by insulating water management from high-level political tensions, interestingly, this same structure paradoxically impedes the expansion of cooperation. Ironically, the very characteristics that make localized cooperation resilient, such as privacy, also prevent water cooperation from expanding into improved international relations or conflict resolution.

Sifundza et al. (2019) examine institutional responses to the 2015/2016 El Niño drought in the Komati catchment of Southern Africa, exposing substantial gaps between scientific capacity and effective drought management despite warning systems and hydrological models. The research demonstrates that the Komati Joint Operations Forum (KJOF), a participatory transboundary governance structure, failed to implement timely water restrictions due to commercial farmers' resistance, resulting in violations of agreements with Mozambique and requiring emergency groundwater extraction measures. The authors advocate for improved institutional coordination through drought task teams at the basin level and improved communication strategies.

Marshall et al. (2017) explore capacity development as a major component of water diplomacy, arguing that effective transboundary water conflict management requires strategic investments in capabilities across all fields. Kimmich and Villamayor-Tomas (2019) use a unique approach to analyzing resource governance by examining networks of "action situations" within irrigation systems.

Madani and Dinar (2012a) use the cooperative game theory as a framework for sustainable common-pool resource management and apply it to a groundwater exploitation issue with different land sizes and pumping costs. The findings demonstrate that while cooperative institutions can provide higher total gains compared to noncooperative management approaches, the stability of these solutions heavily depends on the noncooperative behavioral characteristics of the resource users. Moreover, different cooperative institutions perform better under different user behavioral contexts, so cooperative management institutions appear to be the most efficient approach for upholding CPR sustainability and maximizing long-term benefits.

Using the same case study, Madani and Dinar (2012b) investigate non-cooperative institutions for sustainable common-pool resource management (CPR) through a groundwater exploitation case study, challenging the traditional "tragedy of the commons" by demonstrating that resource users can avoid poor outcomes by modifying their decision-making approaches. The authors introduce six distinct non-cooperative management institutions ranging from ignorant myopic management to smart non-myopic management, in the context of the three-farmer groundwater system over a 50-year planning horizon. By analyzing how different behavioral characteristics and planning horizons affect both individual farmer profits and overall resource sustainability, the authors reveal that long-term planning consistently outperforms short-term approaches, especially when considering externalities, which improves outcomes. They suggest that education about long-term consequences and real-time monitoring can encourage sustainable resource use even in competitive environments.

Keen (2003) examines the challenges facing integrated water management in the South Pacific, using Fiji's capital, Suva, as a case study to illustrate the multifaceted nature of water governance issues in Pacific Island countries. Uniquely, she focuses on the socio-cultural barriers to water management reform in Pacific Island contexts, highlighting how customary land tenure systems and traditional concepts like "vanua", which equates land, water, and human habitats as the same, create distinctive governance challenges not found anywhere else. Although the South Pacific faces similar technical water infrastructure problems as other developing nations, their small scale, unique geographical characteristics, and cultural contexts require fundamentally different institutional approaches to water management. The author argues that effective integrated water management must address technical capacities, infrastructure development, and the complex relationship between traditional governance systems and modern regulatory frameworks, emphasizing the need for culturally sensitive policy approaches and inter-agency coordination to achieve sustainable management.

Mao et al. (2019) explore how China employs a comprehensive institutional framework to achieve SDG 14, focusing on a top-down governmental approach combining formal plans with administrative guidelines to control coastal water pollution. The institutional intervention succeeds through three main mechanisms, which include a Target Responsibility System that links career advancement with environmental performance indicators, a vertical leadership structure, and financial incentives from the government. Uniquely, China focuses on political accountability measures, such as the "veto targets" system, where failure to meet environmental goals diminishes other performance achievements. Although this centralized institutional approach created tangible improvements in coastal water quality, it still faces challenges in inter-sectoral coordination and transboundary pollution management.

Ferguson et al. (2013) analyze Melbourne's transformation from traditional centralized water infrastructure to an integrated hybrid system between 1997 and 2012, primarily driven by the Millennium Drought that generated significant institutional changes across cultural-cognitive, normative, and regulative dimensions. Their research identifies key causal factors, such as shifts in professional beliefs about environmental predictability, development of new technical knowledge, expanded water servicing goals, and improved governance coordination across previously excluded sectors. The authors argue that successful water system transformation requires technological innovation and comprehensive institutional reform.

Empinotti et al. (2019) examine the 2013–2015 water crisis in São Paulo, Brazil, to analyze how governance structures and processes shaped both the effects of the drought and its institutional responses, challenging conventional assumptions that "good governance" fosters water security. The authors explain that the mutualistic commercial relationship between the São Paulo state government and the water utility SABESP influenced the crisis management, prioritizing protecting revenue streams and assets over implementing water conservation measures or other governance methods. By analyzing the evolution of Brazil's water institutional framework, the findings demonstrate how supposedly progressive governance principles, like decentralization, were actually systematically undermined during the crisis by state actors applying supply-led infrastructure solutions instead of demand management.

Mukhtarov et al. (2015) studied the failure of water user associations (WUAs) in Turkey, Azerbaijan, and Uzbekistan to demonstrate how a top-down institutional design neglects local environments, resulting in ineffective reforms in water governance. By understanding how these standardized institutional interventions are implemented nationwide without meaningful engagement or consideration for context, we can see that they are fundamentally misaligned with existing local conditions and practices. As an alternative to conventional top-down approaches, the authors propose an "interactive institutional design" as a collaborative method, emphasizing participatory processes and being adaptive to local contexts.

Ananda and Proctor (2013) examine the effectiveness of collaborative water management approaches through an institutional analysis of the Howard River Catchment in northern Australia, showing how existing institutions can actually constrain collaborative initiatives. Using the Institutional Analysis and Development (IAD) framework, the authors analyze how hierarchies of multiple institutions create structural barriers for collaborative water planning, such as administrative inflexibility and power imbalances. Although collaborative approaches are meant to reduce transaction costs and improve engagement in management, the Howard River Catchment exemplifies that overlaying collaborative processes onto existing institutional hierarchies can actually increase transaction costs and limit participation. Their findings also uniquely support the "Ecology of Games" hypothesis that new collaborative institutions may create additional complexity when imposed upon existing governance structures.

Schnegg and Linke (2015) investigate pastoral communities in northwestern Namibia, understanding why formal sanctions are rarely used despite being clearly defined. Yet, water governance systems still function effectively. The authors find that traditional social relationships, in which members share resources and personal ties, actually create social constraints that prevent the formal assignment of sanctions, but utilize social control instead. This demonstrates how formal relationships damage smaller communities with tight-knit relationships because social hierarchies and respect take precedence over institutional rules. However, they maintain control through gossip, social pressure, and long-term reciprocity expectations, allowing them to succeed without common pool resource management mechanisms.

Chaffin et al. (2016) examine the transition toward adaptive water governance in the Klamath River Basin from 2001 to 2010, highlighting formal relationships between organizations. The results show how the governance network evolved from a hierarchical and government-centered structure to a more polycentric arrangement with multiple branches of power. Although quantitative network metrics alone proved insufficient to identify adaptive governance transitions, the network analysis and qualitative context together explain how institutional relationships can build adaptive capacity through improved communication and collaboration.

Chereni (2007) explores the implementation challenges of IWRM in Zimbabwe's Mazowe Catchment, revealing that the institutional framework suffers from an "institutional pile-up", which is when organizations operate without clearly defined roles, responsibilities, or accountability mechanisms. The results show that ambiguous institutional relationships lead to power struggles and conflicts over water resource decisions, highlighting tensions between traditional and newer authorities regarding power, ownership, and reforms. The author concludes that effective IWRM implementation requires cordial working relationships.

Deng et al. (2016) evaluate China's water environmental management system and its progress toward institutional integration, using Europe's Water Framework Directive (WFD) model as a reference. The authors examine the regulatory, administrative, monitoring, and public participation dimensions of China's water governance in comparison to those of Europe. While China has declared anti-pollution measures, political commitments, and public investments, its water management system remains fragmented because of conflicting interests in government. Incorporating regulatory and administrative coordination, integration, and adoption of a watershed-based management model, China can overcome these institutional challenges through institutional rationalization.

Sokile et al. (2003) analyze the institutional failures preventing effective water management in Tanzania's Rufiji Basin, where population growth and increasing water demands have produced fragmented governance with conflicts and gaps. By identifying that Tanzania's water management institutions operate in isolation, the authors explain how this separates them from effective water management. They propose establishing a comprehensive institutional framework that redistributes responsibilities from centralized state control to participatory water user associations, emphasizing engagement and adaptive governance.

Moss (2004) investigates how the EU Water Framework Directive's implementation in Germany creates opportunities for integrated governance through river basin management approaches. By examining whether river basin management can effectively address institutional coordination issues, the author believes interactive governance and alignment of new frameworks and existing structures can overcome these dilemmas.

Saravanan (2015) utilizes a systems-based analytical framework to examine how multiple actors and institutional rules interact within water governance systems, focusing on a case study from an Indian Himalayan village. The results indicate that water management operates as a complex socio-political process, revealing that effective water governance requires adaptive institutional behaviors instead of rigid policies that fail to account for evolution. As a result, sustainable water management depends on comprehensive frameworks that include both regulatory structures and locally developed institutional arrangements through enhanced infrastructure and capacity-building initiatives.

Grison et al. (2023) assess Integrated Water Resources Management (IWRM) performance across global cities, specifically examining "200 cities in total [that] represent more than 95% of global urban population". By using the City Blueprint Approach lens to measure urban water management effectiveness, the authors identify 24 performance indicators, which include infrastructure quality, wastewater treatment, climate resilience, and governance frameworks. Uniquely, the authors use an innovative statistical estimation model to conclude that government effectiveness deters urban water management success, supporting the institutional perspective that effective governance forms a successful foundation for management. The findings expose global disparities in water management capacity, highlighting substantial deficiencies in progress toward water-related Sustainable Development Goals (SDG), especially in African and Latin American urban areas that severely lack infrastructure and management.

Pahl-Wostl and Knieper (2023) investigate requirements for governance systems to become polycentric, and to effectively tackle water management challenges, applying Qualitative Comparative Analysis (QCA) to data from 26 cases in developed and less developed countries. Results suggest that polycentric governance systems are characterized by high performance in supporting effective coordination compared to fragmented and centralized, not coordinated systems that are characterized by poor performance. In addition, the QCA analyses point to the importance of high institutional capacity for the effectiveness of formal coordination institutions.

Bantider et al. (2023) investigate the different deliberations about water governance and their implications for satisfying the growing demand for water governance in the Ethiopian Central Rift Valley (ECRV). Data were collected through literature surveys and intensive fieldwork, and were analyzed following a discourse analysis and using narrative analysis techniques. The study found that the dominant competing discourses that have greatly influenced water governance in the ECRV focus on decentralization, water-centered development, marketization, land/water degradation, climate change, water scarcity, and weak water governance.

Ávila et al. (2023) address the functionality of the Water Supply and Sanitation services in Brazil, based on the Institutional Analysis and Development (IAD) Framework.

Lai and Zhao (2023) apply the institutional analysis and development framework to identify the actual operating rules of Australia's Murray-Darling Basin (MDB). The results provide insights regarding the key actors in the MDB's institutional structures and their characteristics. These include the position, boundary, choice, aggregation, information, scope, and payoff rules that regulate the water management in the MDB.

Vitale et al. (2023) examine the multi-level governance failures in flood risk management within Naples' Sarno River basin using an IAD framework to understand these gaps. Despite comprehensive flood risk policies established by basin authorities, implementation has been impacted by informal institutions and illegal activity on local decision-making processes. The results expose that illegal construction in high-risk flood zones is facilitated by building amnesties and organized crime, exemplifying how informal rules can undermine formal institutions.

Zinabu et al. (2024) analyze the institutional capacity and water quality modeling challenges in Ethiopia's Awash Basin, focusing on the gap between available data and the practical implementation of in-stream water quality models for pollution control. The authors assessed seven governmental institutions responsible for water quality management, indicating significant deficiencies in database infrastructure, technical expertise, and modeling capabilities that interfere with effective water resource protection. They also evaluate six different water quality models using locally-relevant criteria, ultimately recommending QUAL2KW and INCA as the best for the basin's conditions.

Mwakalila and Muneeni (2025) investigate the effectiveness of local water institutions' management of water resources in the Sanya-Kware sub-catchment in Tanzania.

Meissner et al. (2013) review the literature on water resource management institutions published between 1997 and 2011 shows that scientists are focusing predominantly on catchment management agencies and their institutionalization and organizational functionality. The paper argues that there is much less focus on other water management entities, such as advisory committees, international water management bodies, irrigation boards, the water tribunal, and water user associations. This partial analysis of institutional management of international water management bodies leads to biased conclusions.

Schmeier and Blumstein (2025) investigate the current state of institutionalized cooperation (via treaties and the basin organizations established by treaties) in internationally shared basins. The analysis uses a comprehensive literature review, the Transboundary Freshwater Dispute Database (TFDD) of Oregon State University (with information on international water treaties and river basin organizations). The chapter identifies various forms of conflict and cooperation over international water and the cooperation mechanisms that were developed over time, how they were implemented, and how they were maintained. The analysis addresses the questions of whether such institutions have made a difference in sustainably and cooperatively managing, and how various future global challenges (climate change, population growth) might be addressed by these institutions.

Hjorth and Madani (2023) present a historical overview, from 1945 onwards, of events with possible impact on water management institutions, showing institutional rigidity during that period. The paper explores how an updated knowledge base could serve a quest for sustainable water governance strategies. The main message of this paper is that a persistent failure to learn is an important reason behind the dire state that the water sector is now in. The paper concludes that for a transformation of the institutional superstructure, adaptive water management (AWM), emerges as a prominent way to embark on a necessary, radical transformation of the water governance systems.

Saleth (2018) summarizes nine papers in a special issue that analyzes how water institutions create incentive environments for sustainable water management across various regional and sectoral scales. The papers use a variety of methodological approaches, ranging from econometric analysis and game theory to historical analysis and case studies, explaining topics like water governance, transaction costs of water transfers, and the evolution of water supply institutional structures. The findings display that water governance operates through economic calculus, so institutional performance depends on the relationships between institutional, economic, and political branches. The author also highlights how institutional inconsistency can create contradictory effects and limit the effectiveness of water management organizations across different contexts.

Hoanh and Suhardiman (2014) analyze the evolution of irrigation policy in Vietnam's Mekong River Delta by focusing on irrigation as a sectoral institution. Kibaroglu (2020), as mentioned, explores Turkey's evolving irrigation governance by studying decentralization efforts, with the aim of improving efficiency and cost recovery. Villamayor-Tomas (2018) investigates how Spanish irrigation associations respond to disturbances. Washington-Ottombre and Evans (2019) investigate how the length of tenure among appropriators influences the institutional success of the Mwea Irrigation Scheme (MIS) in Kenya. Cody (2019), explores how internalized cultural norms influence performance and the governance of irrigation systems in the Upper Rio Grande Basin.

Dellapenna et al. (2013) study the future of global water governance, examining the relationship between global water governance structures and their responsiveness to projected water, specifically within the United Nations. The study suggests that for water institutions considering developing governance strategies in order to respond to emerging global water crises, policymakers must first decide what kind of water future they want, for example, how they want to ensure that the two overarching principles of global water governance, access for all and sustainable water resource management, are met.

Gupta and Pahl-Wostl (2013) explore the challenges of global water governance and address them. Mulroy (2017) uses Nebraska's water governance framework to argue that there are benefits to depending on decentralized overlapping power to deal with climate change. This leads to a variety of targeted, localized solutions that foster regulatory experimentation and possibilities for jurisdictional learning.

Grafton et al. (2019) use WGRF to establish water supply and demand convergence to maintain the sustainability of freshwater ecosystem services. The WGRF consists of seven critical strategic considerations regarding water reform and implementation of the framework, which is designed to be flexible for use in any country and integrative regarding the reform research goal, inequities in water allocation, and simple to use or adaptable to many scales and contexts. The seven considerations of WGRF include: "(1) well-defined and publicly available reform objectives; (2) transparency in decision-making and public access to available data; (3) water valuation of uses and non-uses to assess trade-offs and winners and losers; (4) compensation for the marginalized or mitigation for persons who are disadvantaged by reform; (5) reform oversight and "champions"; (6) capacity to deliver; and (7) resilient decision-making that is both beneficial and durable from a broad socio-economic Perspective". The study applies WGRF to the following countries: Murray–Darling Basin (Australia), Rufiji Basin (Tanzania), Colorado Basin (USA and Mexico), and Vietnam. The study's evaluation of WGRF in these five countries proved that the framework was successful. They believe it should be the "core" of water governance reform efforts since it is adaptable enough at the local, basin, and national levels. For the Murray-Darling Basin and Australia, they provide a framework for assessing reform successes and failures, highlighting power imbalances, and, most crucially, providing constructive assistance for adaptive reform management. Second, when applied to Tanzania's Rufiji Basin, the WGRF examines the existing water governance system and, as a result, suggests future water reform options.

Berardo et al. (2013) evaluate the development of several institutions by using the adaptive governance framework applied at the national level. Grigg (2018) evaluates the collective action in multilevel water governance and management in the Western U.S., showing how the nature of water management challenges influences collective action. Wegerich et al. (2014) investigate the evolving dynamics of water governance, specifically in the irrigation sector.

Greitens (2016) study on transboundary governance capacity in the Great Lakes examines AOCs' underlying transboundary governance architecture in terms of "functional intensity, nature of compliance mechanisms, stability and resilience, and legitimacy." According to the findings, transboundary governance requires more centralized enforcement and finance for the AOCs to continue to progress. The findings reveal that AOCs might have considerable transboundary governance flaws in terms of compliance procedures and ideas of stability and resilience, but significant functional intensity and legitimacy. Transboundary governance in the AOCs tends to deteriorate over time, even when early governance solutions are established well with strong stakeholder engagement.

Garrick et al. (2016) utilize the TGC framework to compare responses to non-point contamination in the Great Lakes and transboundary water basins in North America and Australia. "Issue linkages" have been employed in the Columbia, Colorado, and Murray-Darling Basins to address governance capacity gaps by drawing on other water-related concerns, such as fisheries and water scarcity, where governance capacity exists. Issue linkages between nonpoint pollution and salmon recovery improve water quality governance, coordination, and harmonization throughout four main US states, including Oregon, Washington, Idaho, and Montana. In every case, building transboundary governance capacity has necessitated a focus on 'process values,' or the processes used to make decisions and implement nonpoint pollution initiatives. Future studies should look at how features of transboundary governance ability change over time in connection with environmental quality indicators and develop finer-grained indicators to guarantee external validity and allow comparisons between case studies.

Bruns (2017) explores polycentric water governance in Southeast Asia. Gupta and Pahl-Wostl (2013) explore the challenges of global water governance and address them. Gupta and Pahl-Wostl (2013). Garrick et al. (2011) analyze the evolution of institutional frameworks governing environmental water allocation in the Western United States. Grigg (2018) evaluates the collective action in multilevel water governance and management in the Western U.S., showing how the nature of water management challenges influences collective action.

Stoa (2017) looked at water governance in Haiti and found that while local stakeholders are engaged, human and financial resources are inadequate to fulfill statutory duties, according to a capacity evaluation of institutions in northern Haiti. The findings imply that local governments and community institutions should be included in water resource management planning, with national or international assistance augmenting capacity and technical support.

Roßner and Zikos (2018) focus on water governance in Uzbekistan and use an economic-framed field experiment to assess the impacts of regulations on different irrigation groups. Balasubramanya et al. (2018) assess the impact of training on water user association (WUA) staff on the performance of mandated duties of the WUA. Babbitt et al. (2015) explore how a localized and integrated approach to managing the water resources management system operates.

Huang et al. (2009) examine the institutional transformation of water management systems in northern China between 1995 and 2004, revealing that few villages shifted from traditional collective management toward market-oriented reforms, including water user associations (WUAs) and contracting arrangements. Counterintuitively, villages with relatively abundant water resources and complex irrigation infrastructure were more likely to implement reforms than those experiencing acute water scarcity, proposing that institutional change occurs under operational flexibility rather than necessity from crisis. While contracting systems effectively utilized financial incentives for water conservation, WUAs struggled to achieve effective participatory governance because village leadership maintained control, highlighting the gap between reform objectives and implementation outcomes.

Huang et al. (2010) present an empirical analysis of institutional reforms in Northern China, focusing on the replacement of traditional collective water management through water user associations and contracting systems. Drawing on data across three provinces, the results reveal that water user associations meant to be participatory institutions are actually controlled by existing village leadership, limiting farmers' participation. However, the limited democratization outperforms the traditional systems, in which the performance can be measured by improved metrics in fee collection and timely water delivery.

Gopalakrishnan et al. (2005) report the results of a global survey and assessment of the structure, evolution, and performance of water institutions in various settings. Zhang et al. (2014) analyze how evolving market conditions influence the local irrigation institutional frameworks that govern water management in China.

Syed and Choudhury (2018) argue that studying cooperation and conflict at the international water level, addressing multiple scales and multiple stakeholders, is required to reach the degree of cooperation and to ensure the effectiveness of transboundary governance structures. Using the Transboundary Water Interaction Nexus (TWINS) approach applied to the Pakistani part of the Indus, the inclusion of scale allows for the accounting for the complexity of transboundary water governance and the design of effective governing institutions.

Zikos and Roggero (2013) perform an analysis of interview data about Cyprus's water conflicts, arising from the split of the island between two communities. Shah and Narain (2019), dispute the GOI's continued arguments of water crisis, while ignoring the causes of the water crisis. Falk et al. (2019) use experimental games to help better understand coordination challenges and establish institutional capacities in Rajasthan/India. Ménard (2017), aims to identify successful market-based incentives to help promote sustainable watershed practices. Kuhn et al. (2016) use the Lake Naivasha Basin in Kenya's Rift Valley as a hydro-economic system to assess the institutional needs of this system.

Wutich et al. (2017) refer to tensions between institutions and existing cultural norms of justice. The paper proposes a new approach to cross-cultural analysis that can investigate these tensions, which can then assess when local cultural norms are likely to facilitate or impede the acceptance of specific institutions. The approach is implemented, using data collected from respondents from five global sites (in Fiji, Ecuador, Paraguay, New Zealand, and the U.S.) using cultural consensus analysis. Findings suggest evidence of similar cultural norms of justice in water in the following domains: human right to water, water governance, water access, environmental stewardship, aspects of water markets, and aspects of water quality and health.

Franzén et al. (2015) attempt to support the EU Water Framework Directive (WFD), which promotes increased participation of stakeholders. Sosa and Zwarteveen (2014) review literature on water and mining in Peru to assess the effectiveness of institutional instruments for safeguarding the sustainability of water resources and the water-dependent ecosystems. in mining regions. Schnegg and Bollig (2016) examine the ability of water user associations in rural communities in Namibia during drought years to administer water services. Gerlak (2004), studied the Danube River and how to strengthen regional governance bodies. Nyagumbo and Rurinda (2012) evaluate the effectiveness of various water management institutions for smallholder farmers in Zimbabwe. Bettini et al. (2015) analyze governance through the concept of a "scale", challenging the conventional view that it is purely a technical matter. Janssen and Anderies (2013) use a multi-method framework to analyze the strength of small-scale irrigation systems. Shivakoti and Bastakoti (2006) analyze the strength of Montane irrigation systems in Thailand. Berardo et al. (2013) evaluate the development of several institutions by using the adaptive governance framework applied at the national level scale. Villamayor-Tomas (2018) investigates how Spanish irrigation associations respond to disturbances.

Means et al. (2002) warn readers about a water infrastructure crisis in the United States, tracking the drastic increases in estimated replacement costs from almost $138.4 billion in 1997 to nearly $1 trillion by 2000. After decades of declining infrastructure investment, the government has created a maintenance deficit that risks doubling or tripling water rates, with smaller systems being disproportionately impacted due to a limited rate base. By analyzing the political ramifications of rising water costs, the authors predict increased "politicization" of water management because candidates may run campaigns on rate reliefs or open the doors to privatization.

Garrick and Aylward (2012) examine how transaction costs influence the institutional performance of market-based environmental water allocation in the Columbia Basin, a region historically over-allocated for agricultural use. Similarly, Kajisa and Dong (2017) analyze the Indus River to illustrate how cooperation and conflict have evolved over time. Gopalakrishnan et al. (2005) provide results from a global survey assessing the structure, evolution, and performance of water institutions across diverse contexts. Söderbaum (2015) focuses on transboundary water management in the Zambezi River Basin, highlighting institutional dynamics in regional cooperation. Li et al. (2011) investigate the effectiveness of China's water abstraction policies, emphasizing the crucial role of institutional frameworks in fostering well-functioning markets. They conclude that government involvement and institutional robustness become especially important when policy instruments rely on administered prices or artificially created markets. Similarly, Zhang et al. (2014) explore how changing market conditions affect local irrigation institutions governing water management in China, while Yang et al. (2003) analyze China's experimental reforms in water rights and markets, particularly within the irrigation sector.

Muchapondwa et al. (2018) aim to identify successful market-based incentives to help promote sustainable watershed practices. Taylor and Eberhard (2020) present a comprehensive review of the successes and failures of various institutional arrangements to protect the Great Barrier Reef. Söderberg (2016) analyzes Swedish water governance as a case to enhance the understanding of policy implementation in complex governance structures.

3.9.4 Role of Institutions and Economic Analysis

Saleth (2018) studies the institutional economics of water by reviewing nine papers that analyze how water institutions create incentive environments for sustainable water management across various regional and sectoral scales. The papers use a variety of methodological approaches, ranging from econometric analysis and game theory to historical analysis and case studies, explaining topics like water governance, transaction costs of water transfers, and the evolution of water supply institutional structures. The findings display that water governance operates through economic calculus, so institutional performance depends on the relationships between institutional, economic, and political branches. The author also highlights how institutional inconsistency can create contradictory effects and limit the effectiveness of water management organizations across different contexts.

Saleth et al. (2016) examine the global irrigation sector by focusing on the institutional and infrastructural mechanisms required to manage growing agricultural water scarcity. Araral (2009) models the incentive problems in foreign aid, focusing on how moral hazard and aid fungibility interact with incentives.

Mogomotsi et al. (2018) explore Botswana's water supply institutions and allocations, showing how their water scarcity results from physical barriers and outdated institutional frameworks for sustainable management. Shah and Narain (2019), dispute the GOI's continued arguments of water crisis, while ignoring the causes of the water crisis.

Punjabi and Johnson (2019) dissect the complex institutional dynamics of rural and urban water conflicts in India to understand how water governance shapes resource allocation between sectors. By comparing Mumbai and Chennai, they observe how Mumbai uses a historical framework of prior appropriation, allowing urban authorities to control rural water resources, which disadvantages tribal irrigation communities. However, Chennai allocates water depending on the market and riparian rights-based institutions, leading to the commercialization of groundwater, which means the usage of unsustainable extraction practices.

Ménard (2017) aims to identify successful market-based incentives to help promote sustainable watershed practices. Kuhn et al. (2016) use the Lake Naivasha Basin in Kenya's Rift Valley as a hydro-economic system to assess the institutional needs of this system. Binz et al. (2016) investigate the institutional processes of the legitimization of reusable potable water in California. Villamayor-Tomas (2018) investigates how Spanish irrigation associations respond to disturbances.

3.9.5 *Means of Water Management (Integrated Water Resources Management—IWRM, Adaptive Governance Framework)*

Ferguson et al. (2013) analyze Melbourne's transformation from traditional centralized water infrastructure to an integrated hybrid system between 1997 and 2012, primarily driven by the Millennium Drought that generated significant institutional changes across cultural-cognitive, normative, and regulative dimensions. Their research identifies key causal factors, such as shifts in professional beliefs about environmental predictability, development of new technical knowledge, expanded water servicing goals, and improved governance coordination across previously excluded sectors. The authors argue that successful transformation of water systems requires technological innovation and comprehensive institutional reform.

Rak et al. (2019) explore state cooperation on implementing the ballast water management convention in the Adriatic Sea. This paper reveals, based on acquired legislative and institutional data, and taking into account regional marine traffic and environmental conditions, that, via proper Adriatic States' collaboration, the integration of current environmental legislation commitments, as well as a stronger interaction between public institutions from the maritime and environmental sectors, may encourage the execution of ballast water management requirements.

Babbitt et al. (2015) explore how a localized and integrated approach to managing water resources management system operates. Peat et al. (2017) examine how water management agencies can create institutional flexibility to effectively implement adaptive management for aquatic ecosystem restoration. Wan et al. (2018) explore why China has failed to develop comprehensive ballast water management policies to prevent invasive aquatic species. Layzer and Schulman (2013) evaluate the adoption of the IWRM in the U.S., using the Chesapeake Bay Program (CBP) as an example.

Huang et al. (2009) examine the institutional transformation of water management systems in northern China between 1995 and 2004, revealing that few villages shifted from traditional collective management toward market-oriented reforms, including water user associations (WUAs) and contracting arrangements. Counterintuitively, villages with relatively abundant water resources and complex irrigation infrastructure were more likely to implement reforms than those experiencing acute water scarcity, proposing that institutional change occurs under operational flexibility rather than necessity from crisis. While contracting systems effectively utilized financial incentives for water conservation, WUAs struggled to achieve effective participatory governance because village leadership maintained control, highlighting the gap between reform objectives and implementation outcomes.

Huang et al. (2010) present an empirical analysis of institutional reforms in Northern China, focusing on the replacement of traditional collective water management through water user associations and contracting systems. Drawing on data across three provinces, the results reveal that water user associations meant to be participatory institutions are actually controlled by existing village leadership, limiting farmers' participation. However, the limited democratization outperforms the traditional systems, in which the performance can be measured by improved metrics in fee collection and timely water delivery.

Berardo et al. (2013) evaluate the development of several institutions by using the adaptive governance framework applied at the national level. Layzer and Schulman (2013) evaluate the adoption of the IWRM in the U.S., using the Chesapeake Bay Program (CBP) as an example. Gopalakrishnan et al. (2005) report the results of a global survey and assessment of the structure, evolution, and performance of water institutions at various settings. Tortajada and Joshi (2013) describe the development, by the city-state of Singapore, of a comprehensive plan for the overall management of its water resources.

3.9.6 *Global Freshwater Resources and Soft Path Solutions*

Gleick (2003) studied global freshwater resources and observed the soft-path solutions for the twenty-first century on "the construction of massive infrastructure in the form of dams, aqueducts, pipelines, and complex centralized treatment plants to meet human demands." Soft path solutions consider the centralized physical infrastructure that ultimately allows for decreased cost in community-scale systems, distributed and open decision-making, water markets, efficient technology, and environmental protection. One study examines the Interbasin Water Transfer (IWT), which aims to restore the ecology of an intermittent transboundary stream and improve recreational opportunities (Ayun, Israel).

3.9.6.1 Irrigation Development and Technology

Ghimire and Griffin (2014), combine the institutional and political orientation of irrigation districts, favoring irrigation over irrigators. Hoanh and Suhardiman (2014), analyze the evolution of irrigation policy in Vietnam's Mekong River Delta by focusing on irrigation as a sectoral institution.

Kibaroglu (2020) explores Turkey's evolving irrigation governance by studying decentralization efforts, with the aim of improving efficiency and cost recovery. Villamayor-Tomas (2018), investigates how Spanish irrigation associations respond to disturbances. Washington-Ottombre and Evans (2019) investigate how the length of tenure among appropriators influences the institutional success of the Mwea Irrigation Scheme (MIS) in Kenya. Cody (2019), explores how internalized cultural norms influence performance and the governance of irrigation systems in the Upper Rio Grande Basin. Li et al. (2017) analyze how farmers' decisions on irrigated land are influenced by uncertainty in water supply, focusing on the priority of water rights systems. Li and Zhao (2018) investigate how utilizing water-efficient irrigation technologies can unintentionally cause increased water use. Dridi and Khanna (2005) analyze how asymmetric information between farmers and regulators influences water allocation, adoption of irrigation technology, and water trading.

3.9.7 *Water Governance (Domestic and International)*

Dellapenna et al. (2013) studied the future of global water governance, examining the relationship between global water governance structures and their responsiveness to shape projected water, specifically in the United Nations. The study suggests that for water institutions considering developing governance strategies in order to respond to emerging global water crises, policymakers must first decide what kind of water future they want, for example, how they want to ensure that the two overarching principles of global water governance, access for all and sustainable water resource management, are met.

Gupta and Pahl-Wostl (2013) explore the challenges of global water governance and address them. Mulroy (2017) uses Nebraska's water governance framework to argue that there are benefits to depending on decentralized overlapping power to deal with climate change. This leads to a variety of targeted, localized solutions that foster regulatory experimentation and possibilities for jurisdictional learning.

Grafton et al. (2019) use WGRF to establish water supply and demand convergence to maintain the sustainability of freshwater ecosystem services. The WGRF consists of seven critical strategic considerations regarding water reform and implementation of the framework, which is designed to be flexible for use in any country and integrative regarding the reform research goal, inequities in water allocation, and simple to use or adaptable to many scales and contexts. The seven considerations of WGRF include: "(1) well-defined and publicly available reform objectives; (2) transparency in decision-making and public access to available data; (3) water valuation of uses and non-uses to assess trade-offs and winners and losers; (4) compensation for the marginalized or mitigation for persons who are disadvantaged by reform; (5) reform oversight and "champions"; (6) capacity to deliver; and (7) resilient decision-making that is both beneficial and durable from a broad socio-economic Perspective". The study applies WGRF to the following countries: Murray–Darling Basin (Australia), Rufiji Basin (Tanzania), Colorado Basin (USA and Mexico), and Vietnam. The study's evaluation of WGRF in these five countries proved that the framework was successful. They believe it should be the "core" of water governance reform efforts since it is adaptable enough at the local, basin, and national levels. For the Murray-Darling Basin and Australia, they provide a framework for assessing reform successes and failures, highlighting power imbalances, and, most crucially, providing constructive assistance for adaptive reform management. Second, when applied to Tanzania's Rufiji Basin, the WGRF examines the existing water governance system and, as a result, suggests future water reform options.

Berardo et al. (2013) evaluate the development of several institutions by using the adaptive governance framework applied at the national level scale. Grigg (2018) evaluates the collective action in multilevel water governance and management in the Western U.S., showing how the nature of water management challenges influences collective action.

Thomas and Warner (2014) examine river basin multi-stakeholder platforms (MSP) and the practice of good water governance in Afghanistan, examining the chosen pilots for the new models, the water management and conflict resolutions in two sub-basins in Northern Afghanistan, lower Kunduz, and Taloqan, and in two different dry years: 2008 and 2011. The study finds significant gaps between practices and models 7 years after implementing "good" water governance principles in Afghanistan. It is crucial to remember that putting IWRM, RBM, and MSP participation into practice with results that add value to existing processes takes time. The study also suggests that in the post-conflict context of Afghanistan, rigorous implementation of the MSP model of contemporary water management may be "counterproductive." What is gained in terms of 'good governance' compliance by a rigid application of the law may be at the expense of performance.

Wegerich et al. (2014), investigate the evolving dynamics of water governance, specifically in the irrigation sector. Greitens (2016) study on transboundary governance capacity in the Great Lakes examines AOCs' underlying transboundary governance architecture in terms of "functional intensity, nature of compliance mechanisms, stability and resilience, and legitimacy." According to the findings, transboundary governance requires more centralized enforcement and finance for the AOCs to continue to progress. The findings reveal that AOCs might have considerable transboundary governance flaws in terms of compliance procedures and ideas of stability and resilience, but significant functional intensity and legitimacy. Transboundary governance in the AOCs tends to deteriorate over time, even when early governance solutions are established well with strong stakeholder engagement.

Garrick et al. (2016) utilize the TGC framework to compare responses to nonpoint contamination in the Great Lakes and transboundary water basins in North America and Australia. "Issue linkages" have been employed in the Columbia, Colorado, and Murray-Darling Basins to address governance capacity gaps by drawing on other water-related concerns, such as fisheries and water scarcity, where governance capacity exists. Issue linkages between nonpoint pollution and salmon recovery improve water quality governance, coordination, and harmonization throughout four main US states, including Oregon, Washington, Idaho, and Montana. In every case, building transboundary governance capacity has necessitated a focus on 'process values,' or the processes used to make decisions and implement nonpoint pollution initiatives. Future studies should look at how features of transboundary governance ability change over time in connection with environmental quality indicators and develop finer-grained indicators to guarantee external validity and allow comparisons between case studies.

Hoekstra (2011) expresses how the River Basin Approach is not always sufficient. Bruns (2017), explores polycentric water governance in Southeast Asia. Camkin and Neto (2016) use a set of essential water issues to analyze the rights and duties of diverse players in water governance.

Gupta and Pahl-Wostl (2013) explore the challenges of global water governance and address whether water challenges are global challenges and what type of governance and water resource management would be appropriate. The article suggests that the worldwide water governance now overcomes management at the river basin level.

Garrick et al. (2011) analyze the evolution of institutional frameworks governing environmental water allocation in the Western United States. Grigg (2018) evaluates the collective action in multilevel water governance and management in the Western U.S., showing how the nature of water management challenges influences collective action. Hoekstra, Chapagain [15], focus on reducing the water footprint of our production and consumption habits.

Stoa (2017) looked at water governance in Haiti and found that while local stakeholders are engaged, human and financial resources are inadequate to fulfill statutory duties, according to a capacity evaluation of institutions in northern Haiti. The findings imply that local governments and community institutions should be included in water resource management planning, with national or international assistance augmenting capacity and technical support.

Roßner and Zikos (2018) focus on water governance in Uzbekistan and used an economic-framed field experiment to assess the impacts of regulations on different irrigation groups. Bakker et al. (2008) analyze Jakarta's urban water supply systems by understanding failures in governance, highlighting how both institutional structures and ownership provide access to water for poor households. Gerlak (2017) emphasizes the importance of participation and engagement of stakeholders in river basin institutions. Balasubramanya et al. (2018) assess the impact of training on water user association (WUA) staff on the performance of mandated duties of the WUA. And Syed and Choudhury (2018) study the Indus River to demonstrate the assimilation of cooperation and conflict over time.

3.9.8 Water Footprint

Makate et al. (2018), review the water footprint methodology to increase stakeholder engagement.

3.9.9 Institutional Fit

Zikos and Roggero (2013) review formal and informal water institutions in the Euphrates-Tigris Region, including Turkey, Syria, and Iraq. Babbitt et al. (2015) explore how a localized and integrated approach to managing the water resources management system operates. Shah and Narain (2019) dispute the GOI's continued arguments of water crisis, while ignoring the causes of the water crisis.

Falk et al. (2019) uses experimental games to help better understand coordination challenges and establish institutional capacities in Rajasthan/India. Ménard (2017) aims to identify successful market-based incentives to help promote sustainable watershed practices. Kuhn et al. (2016) use the Lake Naivasha Basin in Kenya's Rift Valley as a hydro-economic system to assess the institutional needs of this system.

Wutich et al. (2017) refer to tensions between institutions and existing cultural norms of justice. The paper proposes a new approach to cross-cultural analysis that can investigate these tensions, which can then assess when local cultural norms are likely to facilitate or impede the acceptance of specific institutions. The approach is implemented by applying cultural consensus analysis to data collected from respondents at five global sites (Fiji, Ecuador, Paraguay, New Zealand, and the U.S.). Findings suggest evidence of similar cultural norms of justice in water in the following domains: human right to water, water governance, water access, environmental stewardship, aspects of water markets, and aspects of water quality and health.

3.9.10 Institutional Development and Regulation

Franzén et al. (2015) attempt at supporting the EU Water Framework Directive (WFD), which promotes increased participation of stakeholders. Sosa and Zwarteveen (2014) review literature on water and mining in Peru to assess the effectiveness of institutional instruments for safeguarding the sustainability of water resources and the water-dependent ecosystems. in mining regions. Schnegg and Bollig (2016) examine the ability of water user associations in rural communities in Namibia during drought years to administer water services.

3.9.11 Institutional Frameworks (River Basin, Smallholder Farmers, Groundwater)

Gerlak (2004) study the Danube River and how to strengthen regional governance bodies. Nyagumbo and Rurinda (2012) evaluate the effectiveness of various water management institutions for smallholder farmers in Zimbabwe. Bettini et al. (2015) analyze governance through the concept of a "scale", challenging the conventional view that it is purely a technical matter. Ananda and Aheeyar (2020) present a comprehensive review of the successes and failures of various institutional arrangements to protect the Great Barrier Reef.

3.9.12 Small Scale Irrigation Systems (Residential, Urban, Minor Irrigation Systems)

Janssen and Anderies (2013) use a multi-method framework to analyze the strength of small-scale irrigation systems. Shivakoti and Bastakoti (2006) analyze the strength of Montane irrigation systems in Thailand. Tiger et al. (2011) study the implications of residential irrigation metering on water demand and consumer expenditures in North Carolina. Cody (2019) explores how internalized cultural norms influence performance and the governance of irrigation systems in the Upper Rio Grande Basin. Hayden and Tsvetanov (2019) investigate irrigation restrictions in California and the effectiveness of the outdoor irrigation water restrictions. Behera and Mishra (2018) evaluate the performance of the long-known water management Panchayat institution in India.

3.10   Miscellaneous (Undefined Institutional Premises)

There are nine papers that were not allocated to the specific bins identified above. Programme (2009) refers to the understanding that policy and legal frameworks are necessary to develop, carry out, and enforce the rules and regulations that govern water use and protect that resource. Effective implementation and enforcement of water laws and regulations require an adequate institutional and governance framework—one that is legitimate, transparent and participatory and that has proper safeguards against corruption. They also highlight the importance of supporting institutional development to prepare institutions to deal with current and future challenges. Identifying the role of institutions in considering the influence of water law, both formal and customary, including regulations in other sectors that influence the management of water resources. Importance of involving stakeholders and ensuring accountability in planning, implementation, and management, as well as building trust within the water and related sectors, and fighting corruption and mismanagement. Finally, they advocate for having inclusive institutional structures in place for multi-stakeholder dialogue and cooperation as a prerequisite for ensuring equitable access to sustainable water supply and sanitation services. Government itself is not sufficient for 'providing' water supply and sanitation services to all citizens, especially in low-income countries. Creating coherence between the various institutional levels is essential to ensure that policies are impactful. The report highlights the role of non-governmental organizations (NGOs) in expressing the opinions of civil society and promoting the public's active participation (Heckman et al. 1997).

Saleth and Dinar (2004) develop a quantitative framework to express the structural and functional linkages within institutional systems and assess their performance implications and strategic importance for promoting institutional reforms. The publication develops an analytical framework for specifying alternative models of institution–performance interaction within the water sector under different assumptions concerning institutional linkages and their structural properties. The publicatio concludes by explaining the policy roles of such institutional linkages and the implementation principles that are useful to overcome the technical and political economy constraints hindering institutional reforms.

Saravanan (2015) examines the roles of water-related agents and their negotiation powers as linked with the institutions through an ethnographic method of long-time observation. Different types of agents were observed: goal-oriented agents, agents maintaining positions, opportunistic, reactive, and supportive agents. Each of these agents integrates institutions based on their past experience and own task rationality to display their power in bringing about institutional change. The paper argues that applying this new institutionalism offers insights into three significant areas of water management and help designing and crafting of institutions to facilitate desired change.

Klassert et al. (2023) present a rigorous coupled human–natural system analysis of rural-to-urban tanker water market supply and demand in Jordan. This type of market in a typical institution in water scarce countries with illegal water abstractions. These markets supply 15% of all drinking water at high prices, account for 52% of all urban water revenue and constrain the public supply agencies ability to recover costs. The paper projects that household reliance on tanker water will grow nearly threefold by 2050 under population growth and climate change.

Iyiola et al. (2024) highlight the importance of coordination between indigenous and modern institutions of sustainable water use to agricultural evolution and the processes, and measures involved. The paper alludes to the process of agricultural development and intensification in Africa that has increased in the past half-decade and is projected to increase further. Modern and improved irrigation water usage have been observed, but it needs inputs from and coordination among the stakeholders, government sector, agricultural extension, and agencies in order to be effective in agricultural growth. The paper identifies that with suitable knowledge about modern approaches to water use for agricultural development such as climate-smart agriculture, it can be used to address sustainability issues as well as to drive the economic growth of the Africa region.

Wheeler et al. (2024) reviews the sustainability challenges in Australia's Murray-Darling Basin (MDB). Facing difficulties from changing and coordinating water management across multiple states, the MDB experiences growing difficulties in sustainable water supply. One of the alternatives is a complete Federal takeover of water resources through a constitutional amendment. The paper investigates the Australian public's desire for such institutional change. Findings suggest a relatively small (40%) support for a Commonwealth takeover of water resources in the MDB. Berge and Torsteinsen (2023), explore whether increasing the governance ability of a water supply agency may influence governance control.

Apio et al. (2025) use meta-regression analysis to investigate the empirical literature on the performance of water institutions. The basic question is whether institutional reform culminated in positive outcomes. The paper synthesizes and quantifies the overall water institution-performance effects, using data from 23 original published studies. The different statistical approaches confirm the presence of a publication selection bias that favors the positive impact of water institutions on performance. Once this bias is corrected, evidence of a genuine empirical effect of water institutions on performance is evident.

Dolan et al. (2021) conduct a global-to-basin-scale exploratory analysis of potential water scarcity impacts by linking a global human-Earth system model, a global hydrologic model, and a metric for the loss of economic surplus due to resource shortages. Findings suggest that, based on scenario assumptions, major hydrologic basins can experience strongly positive or strongly negative economic impacts due to global trade dynamics and market adaptation ability to regional scarcity. In many cases, the institution of market adaptation profoundly magnifies economic uncertainty relative to hydrologic uncertainty.

O'Donnell (2023) refers to the water theft and the injustice of Indigenous Peoples by new settlers using examples from Aotearoa New Zealand, the USA, Canada, and Australia. The analysis of the cases suggests that acknowledging and challenging the assumption of 'aqua nullius'—that indigenous people have no historical or legal rights to water—creates pathways for reform, which enables pluralist water laws and water governance models that improve both legitimacy and sustainability of settler state water governance. This approach could be of special relevance in the case of the Indian tribes' water rights discourse in the Colorado River Basin.

Chapter 4
Conclusion and Further Work Needed

Abstract This Chapter provides a snapshot of the institutional changes observed over time and space in the review. It also provides a general conclusion on the status and overall performance of water institutions (Sect. 4.1), and specific conclusions for the individual bins (Sect. 4.2). In addition, the chapter provides the authors' suggestions regarding future work on water institutions (Sect. 4.3).

4.1 Institutional Changes over Time and Space

Recent water-related episodes leave little doubt that the world faces an ongoing and escalating problem of water scarcity, affecting quantity, quality of water availability, equity, and political consequences of their allocation and use. This trend is evident in many countries, both developing and developed. While the nature and extent of such problems differ across countries, in many countries, water scarcity, either in terms of quantity or quality, or both, is the result of inefficient use and mismanagement of water resources, which lasts too long (Saleth and Dinar 2004).

This ongoing water scarcity crisis is further exacerbated by climate change, population growth, and lifestyle changes worldwide. The water scarcity crisis has several dimensions: physical, economic, policy, and institutional. The first two dimensions—physical and economic may be addressed by policies and institutional reforms. However, it is evident that policy instruments developed in the era of plenty have not been adapted to the era of scarcity, and that existing water institutional reforms are limited in addressing the new set of problems related more to resource allocation and management than to water resources development.

The paradigm shifting from water as a free good to water as an economic and social good means that the regulatory policies of water pricing and the considerations applied to water project selection need a change that reflects such a shift. This opens the door for new water allocation institutions and conflict-resolution mechanisms to be created or adapted. New legal and policy domains need to be developed. Allocation and conflict-resolution mechanisms need to be created, strengthened, or restructured. Means of decentralization of water management could replace central

A. Dinar et al., *Water Institutions and their Performance*, SpringerBriefs in Water Science and Technology, https://doi.org/10.1007/978-3-032-24649-3_4

government management mechanisms. Water user associations, nongovernmental agencies, women, and local advocacy groups should lead information and know-how transfer. And several other institutional features can emerge and introduce incentives to use water more efficiently.

This set of water institutions mentioned above, which may help improve water management efficiency, is only a partial one. The comprehensive review of water institutions and their performance over the past three decades in this book suggests that the potential of water institutions to improve water use efficiency is substantial. The following paragraphs summarize the conclusions that can be drawn from the review across various world regions, countries, sectors, and institutional frameworks. The bin methodology we used enables us to compare institutional performance across different layers and scales, and to appreciate the intricacy of institutional arrangements and their sensitivity to physical, social, economic, and political/constitutional contexts.

4.2 General Conclusion and Bin-Level Conclusions

The review has identified different scales or layers, including global, cross-country/comparative, national, subnational/regional/river-basin, and local/community/utility. The institutions discussed at each scale are not expected to perform with the same success (given all other conditions are similar) at other scales. Institutions at the global scale, for example, incorporate considerations beyond water. They also address issues related to the trade of water-consuming goods, politics, social norms, cultural, and constitutional differences that make it harder to compare institutional design and performance across jurisdictions.

The review has also distinguished among institutional analyses at sectoral levels. The main sectors covered in the literature reviewed include irrigated agriculture, residential (urban/municipal), industrial, and environmental. Common institutional themes recurring in the papers include institutional performance in environmental water allocation, water management, integrated regional water management (IRWM), incentives (e.g., prices, taxes), water trade (inter-sectoral and inter-regional), and water rights. We focus here on two aspects (that may be connected), which characterize all sectors in this review: (1) the problematic nature of design and implement pricing schemes that send the right signal to water users, and be followed, and (2) the failure of a central management system (government) to efficiently manage the resource. These two aspects give rise to the call for decentralization of water resource management, not as a silver bullet without its own issues.

Since the water sector often operates in conjunction with other sectors (e.g., water-energy-food), synchronizing the use of water across multiple integrated sectors is crucial. Multi-sector water institutions are considered to be impactful. Therefore, an economy-wide analysis, incorporating major water-using and water-producing sectors, could amplify the effects of both successful and/or failed institutions. One of the most important aspects supporting the development of multisector institutions is a country's water security or food security, both of which are particularly significant in developing countries, especially in countries that share water resources (rivers, groundwater aquifers, lakes).

Based on the premise that strong and well performing institutions need a more stable economic structure, a higher level of human capital, to name a few parameters that are usually considered fundamental for strong institutions, one would expect that developed countries would often have more developed water institutions and policies that shape their water systems; however, this has not been reflected in the review that we conducted. Our observations suggest that due to the multiple parameters that affect the well performance of institutions, especially in the water sector, the distinction between developed and developing environments does not solely affect the institutional performance. Examples in the literature suggest that, depending on exogenous conditions and forces, institutions in developed countries may underperform, whereas institutions in developing countries may perform better. Several of the papers we reviewed highlighted the importance of indigenous institutions, which are highly characteristic of the developing world.

As water scarcity worsens, users introduce new (and sometimes manufactured) types of water and develop technologies as well as institutions to allow their proper use. Among the types of water used in different settings, one can find the 'traditional' surface water and groundwater, but in recent years, the introduction of wastewater, desalinated seawater, and desalinated brackish water, and the conjunctive use of all or part of these types, in different locations worldwide. The reviewed studies have revealed the variability in the institutions developed and applied, where each institution follows different frameworks, parameters, and criteria in various regions. For example, the transition from using surface water only to treated wastewater requires adjustments in the compensation tradeoff between these two types, which is also associated with negotiations between the regulator and potential users. Another difficult aspect of the use of treated wastewater is the health regulations and the institutions that support them (e.g., PFAS, Nutrients). Finally, because manufactured water (desalinated and treated) often requires substantial financial investment, a major challenge in its use is the institution that allocates the joint cost among users and producers. The three challenges we alluded to are only part of the difficulties faced by institutions aimed at allowing the use of different types of water.

One of the most challenging aspects of water management is the ability of existing institutions to handle the management of international (or transboundary) water that is shared by several riparian states (river basins, aquifers, lakes). The literature highlights the role of agreements (treaties) grounded in the principles of international water law, which have been adopted by most countries worldwide. A treaty comprises several institutions that together enable the reasonable management of shared water. Treaties allow for institutions such as river basin agencies, compensation mechanisms, monitoring, and data sharing. But treaty institutions are also susceptible to geographic, power relations, and non-water-related conflicts among the riparian states, which make them, at times, and especially under climatic uncertainty, dysfunctional. Several works addressed the importance and value of informal institutions in the case of international water. Informal institutions, such as the indigenous institutions that we addressed earlier, can play an important supporting role in the functionality of the formal institutions (treaties and their derivatives) in managing complex international water management situations.

With the increasing realization of the impact of climate change, the literature responds with increased attention. As some regions are projected to experience increased impacts of climate change, including frequent droughts, floods, wildfires, and severe heat waves, there is likely to be greater emphasis on institutional capacity to address climate change in these regions. However, from Fig. 3.2, it seems that the work has been concentrated "under the street light pole". Institutions to help the water sector adapt to climate change are crucial because they can affect water availability, quality, and use in many ways, such as flooding, water shortages, drinking water, water for sanitation, industry, and crop irrigation. The review suggests very few institutional interventions, and mainly the establishment of intra- and inter-sectoral and regional water transfer opportunities (such as water trade). While this option is indeed economically superior and efficient, it requires adequate software (including water rights, local, and national laws) and hardware (such as canals, reservoirs, and pipes) to connect supply and demand nodes. These requirements are not always in place, which renders the water trade largely irrelevant in many areas, particularly in developing countries. Agricultural extension is a very effective institution that transfers knowledge about climate adaptation to farmers. Unfortunately, our review identified only one study that estimates the role of agricultural extension as an institution to alleviate farmers' ability to cope with climate change.

The success of water institutions depends on the interventions employed and their capacity to withstand unforeseen events, such as external shocks or conflicts among constituents. There are numerous intervention strategies and methods that are specific to different sectoral institutions (irrigation, residential, industrial, environmental, and international water). Not all intervention methods share the same purpose and are not necessarily the best fit for all. The most common themes present in this review include the role of institutions and economics, IWRM, environmental cooperation, global freshwater resources, soft path solutions, irrigation development and technology, water governance, and water user associations. We lump them under a broader group classifications: Legal (water rights, water quotas, standards); Policy (pricing, subsidies [direct and indirect], taxes, trade); Organizational (user organization, decentralization, enhancing capacity); Integrated water resources management (IWRM); Global freshwater resources and soft path solutions; Water governance; Global freshwater resources and soft path solutions; with several subgroups as well. The overall takeaway from the papers in this bin is that water institutions are highly fragile and sensitive to the complexity of the water system they are intended to regulate. In addition, water institutions may not withstand uncertainty (both external and internal). Therefore, the simpler the institutional means developed for a water system, the better the chances that they will sustain unexpected situations.

4.3 Suggested Future Work

The review of 275 papers published over three decades presents a unique opportunity to track developments in institutional analysis over time and to highlight areas that require further attention. The institutional analysis in the work reviewed in this book is mainly descriptive (qualitative), although we can detect a more quantitative analysis towards the second half of the period (after 2010) we review. Future research in water institutions would benefit and be more effectively communicated to policymakers if it includes a clear ranking and a quantitative (either monetary or index-based) assessment of the performance of the institutions at stake. Another observation from the many works we reviewed is that relatively few works have moved beyond describing the costs associated with implementing institutional means. Such a cost may be prohibitive and prevent the implementation of the suggested institution under certain circumstances. Future work would benefit from including the transaction costs associated with each institutional means. Then, a comprehensive analysis that considers both the benefits and costs of the institutional means could provide a foundation for a complete recommendation for one institutional line that is superior to the others.

A final suggestion, based on our observations from the reviewed studies, is to increase the focus on institutions that address equity-consequences of water mismanagement and create a three-pod model to allow the assessment of the tradeoff among institutional implementation cost, institutional implementation efficiency outcome, and effects on equity at various scales and water scarcity levels, such as: rural-urban, sub-regions, interstate, and moderate and severe water scarcity.

References

Abbas F, Al-Naemi S, Farooque AA, Phillips M (2023) A review on the water dimensions, security, and governance for two distinct regions. Water 15(1):208

Agarwal S, Araral E, Fan M, Qin Y, Zheng H (2023) The effects of policy announcement, prices and subsidies on water consumption. Nat Water 1(2):176–186

Akron A, Ghermandi A, Dayan T, Hershkovitz Y (2017) Interbasin water transfer for the rehabilitation of a transboundary Mediterranean stream: an economic analysis. J Environ Manag 202:276–286

Al-Saidi M (2025) Institutions in the Nile Basin: extending the present state of transboundary affairs. In: World scientific handbook of transboundary water management: science, economics, policy and politics: volume 3: the role of formal and informal institutions in managing transboundary basins. World Scientific, Singapore, pp 251–276

Al-Saidi M, Hefny A (2018) Institutional arrangements for beneficial regional cooperation on water, energy and food priority issues in the Eastern Nile Basin. J Hydrol 562:821–831

Altingoz M, Ali SH (2019) Environmental cooperation in conflict zones: riparian infrastructure at the Armenian–Turkish border. J Environ Dev 28(3):309–335

Ananda J, Aheeyar M (2020) An evaluation of groundwater institutions in India: a property rights perspective. Environ Dev Sustain 22(6):5731–5749

Ananda J, Proctor W (2013) Collaborative approaches to water management and planning: an institutional perspective. Ecol Econ 86:97–106

Apio AT, Thiam DR, Dinar A (2025) A meta-analysis of water institutions and their performance: implications for water resource management. Water Resour Manag 39(2):907–938

Araral E (2009) The strategic games that donors and bureaucrats play: an institutional rational choice analysis. J Public Adm Res Theory 19(4):853–871

Araral E (2010) Improving effectiveness and efficiency in the water sector: institutions, infrastructure and indicators. Water Policy 12(S1):1–7

Araral E, Ratra S (2016) Water governance in India and China: comparison of water law, policy and administration. Water Policy 18(S1):14–31

Araral E, Wang Y (2013) Water demand management: review of literature and comparison in South-East Asia. Int J Water Resour Dev 29(3):434–450

Ashraf T (2025) River management in the GBM Basin: a comparative analysis of formal and informal water management institutions in India and Bangladesh. In: World scientific handbook of transboundary water management: science, economics, policy and politics: volume 3: the role of formal and informal institutions in managing transboundary basins. World Scientific, Singapore, pp 277–303

Ávila NRA, De Carvalho BE, de Moraes Cordeiro Netto O (2023) Water and sanitation services in Brazil: evaluating the regulatory governance based on the institutional analysis and development framework. Water Policy 25(7):639–655

Azhoni A, Holman I, Jude S (2017a) Adapting water management to climate change: institutional involvement, inter-institutional networks and barriers in India. Glob Environ Chang 44:144–157

Azhoni A, Holman I, Jude S (2017b) Contextual and interdependent causes of climate change adaptation barriers: insights from water management institutions in Himachal Pradesh, India. Sci Total Environ 576:817–828

Babbitt CH, Burbach M, Pennisi L (2015) A mixed-methods approach to assessing success in transitioning water management institutions: a case study of the Platte River Basin, Nebraska. Ecol Soc 20(1)

Bakker K, Kooy M, Shofiani NE, Martijn E-J (2008) Governance failure: rethinking the institutional dimensions of urban water supply to poor households. World Dev 36(10):1891–1915

Balasubramanya S, Price JPG, Horbulyk TM (2018) Impacts assessments without true baselines: assessing the relative effects of training on the performance of water user associations in Southern Tajikistan. Water Econ Policy 4(3):1850007

Bantider A, Tadesse B, Mersha AN, Zeleke G, Alemayehu T, Nagheeby M, Amezaga J (2023) Voices in shaping water governance: exploring discourses in the Central Rift Valley, Ethiopia. Water 15(4):803

Barbier E (2019) 5. Reforming governance and institutions. In: The water paradox. Yale University Press, New Haven, pp 110–133

Bastakoti RC, Shivakoti GP (2012) Rules and collective action: an institutional analysis of the performance of irrigation systems in Nepal. J Inst Econ 8(2):225–246

Behera B, Mishra P (2018) Democratic local institutions for sustainable management and use of minor irrigation systems: experience of Pani Panchayats in Odisha, India. Water Econ Policy 4(3):1850010

Berardo R, Meyer M, Olivier T (2013) Adaptive governance and integrated water resources management in Argentina. Int J Water Gov 1(3–4):219–236

Berge DM, Torsteinsen H (2023) Governance challenges of different institutional logics and modes of organising: a Norwegian case study of municipal water supply. Local Gov Stud 49(3):471–491

Bettini Y, Brown RR, de Haan FJ, Farrelly M (2015) Understanding institutional capacity for urban water transitions. Technol Forecast Soc Chang 94:65–79

Bilalova S, Newig J, Tremblay-Lévesque L-C, Roux J, Herron C, Crane S (2023) Pathways to water sustainability? A global study assessing the benefits of integrated water resources management. J Environ Manag 343:118179

Binz C, Harris-Lovett S, Kiparsky M, Sedlak DL, Truffer B (2016) The thorny road to technology legitimation—institutional work for potable water reuse in California. Technol Forecast Soc Chang 103:249–263

Bitterman P, Koliba C, Singer A (2023) A network perspective on multi-scale water governance in the Lake Champlain Basin, Vermont. Ecol Soc 28(1):44

Bolognesi T, Pflieger G (2019) The coherence(s) of institutional resource regimes: typology and assessments from the case of water supply management. Environ Sci Pol 99:17–28

Brent DA (2017) The value of heterogeneous property rights and the costs of water volatility. Am J Agric Econ 99(1):73–102

Brodnik C, Brown R, Cocklin C (2017) The institutional dynamics of stability and practice change: the urban water management sector of Australia (1970–2015). Water Resour Manag 31(7):2299–2314

Bromley DW, Anderson G (2018) Does water governance matter? Water Econ Policy 4(3):1750002

Bruns B (2017) Challenges of polycentric water governance in Southeast Asia: awkward facts, missing mechanisms, and working with institutional diversity. In: Redefining diversity & dynamics of natural resources management in Asia, Volume 1. Elsevier, Amsterdam, pp 55–66

Bulengela G (2024) Rethinking the role of institutions in water resource governance in Tanzania: what is still missing? Eur J Theor Appl Sci 2:259–268

Byrnes J (2013) A short institutional and regulatory history of the Australian urban water sector. Util Policy 24:11–19

Camkin J, Neto S (2016) Roles, rights, and responsibilities in water governance: reframing the water governance debate. World Aff 179(3):82–112

Chaffin BC, Garmestani AS, Gosnell H, Craig RK (2016) Institutional networks and adaptive water governance in the Klamath River Basin, USA. Environ Sci Policy 57:112–121

Chang H-H, Boisvert RN, Hung L-Y (2010) Land subsidence, production efficiency, and the decision of aquacultural firms in Taiwan to discontinue production. Ecol Econ 69(12):2448–2456

Chen N, Hong H, Gao X (2021) Securing drinking water resources for a coastal city under global change: scientific and institutional perspectives. Ocean Coast Manag 207:104427

Chereni A (2007) The problem of institutional fit in integrated water resources management: a case of Zimbabwe's Mazowe catchment. Phys Chem Earth Parts A/B/C 32(15–18):1246–1256

Cobourn KM (2015) Externalities and simultaneity in surface water-groundwater systems: challenges for water rights institutions. Am J Agric Econ 97(3):786–808

Cody K (2019) The evolution of norms and their influence on performance among self-governing irrigation systems in the Southwestern United States. Int J Commons 13(1):578–608

Colby B, Reed-Spitzer Z (2024) Evaluating institutional success: regional water agreements with tribal nations. J Nat Resour Policy Res 10(2):79–102

Colebatch HK (2006) Governing the use of water: the institutional context. Desalination 187(1–3):17–27

Crespo D, Albiac J, Kahil T, Esteban E, Baccour S (2019) Tradeoffs between water uses and environmental flows: a hydroeconomic analysis in the Ebro Basin. Water Resour Manag 33(7):2301–2317

da Silva LPB, dos Santos Brito ÁG, Ribeiro WC (2025) Transboundary water governance institutions in La Plata Basin: the Quaraí Sub-Basin and the role of rice farmers. In: World scientific handbook of transboundary water management: science, economics, policy and politics: volume 3: the role of formal and informal institutions in managing transboundary basins. World Scientific, Singapore, pp 223–250

Dadson S, Hall JW, Garrick D, Sadoff C, Grey D, Whittington D (2017) Water security, risk, and economic growth: insights from a dynamical systems model. Water Resour Res 53(8):6425–6438

Dellapenna JW (2025) Governing the north American Great Lakes across national and subnational borders. In: World scientific handbook of transboundary water management: science, economics, policy and politics: volume 3: the role of formal and informal institutions in managing transboundary basins. World Scientific, Singapore, pp 141–178

Dellapenna JW, Gupta J, Li W, Schmidt F (2013) Thinking about the future of global water governance. Ecol Soc 18(3):28

Deng Y, Brombal D, Farah PD, Moriggi A, Critto A, Zhou Y, Marcomini A (2016) China's water environmental management towards institutional integration. A review of current progress and constraints Vis-a-Vis the European experience. J Clean Prod 113:285–298

Dharmaratna D (2011) Demand, supply and welfare aspects of pipe-borne water in Sri Lanka. Cambridge Scholars Publishing, Newcastle upon Tyne

Diao X, Roe T (2003) Can a water market avert the "double-whammy" of trade reform and lead to a "win–win" outcome? J Environ Econ Manag 45(3):708–723

Dinar A (2016) Dealing with water scarcity: need for economy-wide considerations and institutions. Choices 31(3):1–7

Dinar S, Dinar A (2016) International water scarcity and variability. University of California Press, Oakland

Dinar A, Tsur Y (2014) Water scarcity and water institutions. In: Routledge handbook of water economics and institutions. Routledge, Milton Park, pp 234–251

Dolan F, Lamontagne J, Link R, Hejazi M, Reed P, Edmonds J (2021) Evaluating the economic impact of water scarcity in a changing world. Nat Commun 12(1):1915

Dridi C, Khanna M (2005) Irrigation technology adoption and gains from water trading under asymmetric information. Am J Agric Econ 87(2):289–301

Easter KW, McCann LMJ (2010) Nested institutions and the need to improve international water institutions. Water Policy 12(4):500–516

Edwards EC, Guilfoos T (2021) The economics of groundwater governance institutions across the globe. Appl Econ Perspect Policy 43(4):1571–1594

Eisenack K (2016) Institutional adaptation to cooling water scarcity for thermoelectric power generation under global warming. Ecol Econ 124:153–163

Emerick K, Lueck D (2015) Economic organization and the structure of water transactions. J Agric Resour Econ 40:347–364

Emmerson C (2011) Water: worlds of water. The World Today 67(7):4–6

Empinotti VL, Budds J, Aversa M (2019) Governance and water security: the role of the water institutional framework in the 2013–15 water crisis in São Paulo, Brazil. Geoforum 98:46–54

Falk T, Kumar S, Srigiri S (2019) Experimental games for developing institutional capacity to manage common water infrastructure in India. Agric Water Manag 221:260–269

Ferguson BC, Brown RR, Frantzeskaki N, de Haan FJ, Deletic A (2013) The enabling institutional context for integrated water management: lessons from Melbourne. Water Res 47(20):7300–7314

Fernandez L (2013) Transboundary water institutions in action. Water Resour Econ 1:20–35

Fowler LB, Swain A, Orozco GG, Attamah S, Rubio AMB, Vitisia B (2025) Transboundary water organizations: mechanisms and challenges for managing evolving disputes. In: World scientific handbook of transboundary water management: science, economics, policy and politics: volume 3: the role of formal and informal institutions in managing transboundary basins. World Scientific, Singapore, pp 97–139

Franzén F, Hammer M, Balfors B (2015) Institutional development for stakeholder participation in local water management—an analysis of two Swedish catchments. Land Use Policy 43:217–227

Friedman KB (2016) Institutions and transboundary governance capacity (TGC) in the Arctic: insights from the TGC framework. Int J Water Gov 4:133–154

Fuenfschilling L, Truffer B (2016) The interplay of institutions, actors and technologies in socio-technical systems—an analysis of transformations in the Australian urban water sector. Technol Forecast Soc Chang 103:298–312

Gaden M (2016) Cross-border Great Lakes fishery management: achieving transboundary governance capacity through a non-binding agreement. Int J Water Gov 4:53–72

Gani A, Scrimgeour F (2014) Modeling governance and water pollution using the institutional ecological economic framework. Econ Model 42:363–372

Garrick D, Aylward B (2012) Transaction costs and institutional performance in market-based environmental water allocation. Land Econ 88(3):536–560

Garrick D, Lane-Miller C, McCoy AL (2011) Institutional innovations to govern environmental water in the Western United States: lessons for Australia's Murray–Darling Basin. Econ Pap A J Appl Econ Policy 30(2):167–184

Garrick DE, Krantzberg G, Jetoo S (2016) Building transboundary water governance capacity for non-point pollution: a comparison of Australia and North America. Int J Water Gov 4:111–132

Garrido A (2007) Designing water markets for unstable climatic conditions: learning from experimental economics. Rev Agric Econ 29(3):520–530

Gerlak AK (2004) Strengthening river basin institutions: the global environment facility and the Danube River Basin. Water Resour Res 40(8):W08S08

Gerlak AK (2017) Regional water institutions and participation in water governance: the Colorado River Delta as an exception to the rule? J Southwest 59(1):184–203

Gharib AA, Arabi M, Goemans C, Manning DT, Maas A (2024) Integrated water management under different water rights institutions and population patterns: methodology and application. Water Resour Res 60(11):e2024WR037196

Ghimire N, Griffin RC (2014) The water transfer effects of alternative irrigation institutions. Am J Agric Econ 96(4):970–990

Gleick PH (2002) Water management: soft water paths. Nature 418(6896):373

Gleick PH (2003) Global freshwater resources: soft-path solutions for the 21st century. Science 302(5650):1524–1528

Gohar AA, Ward FA (2010) Gains from expanded irrigation water trading in Egypt: an integrated basin approach. Ecol Econ 69(12):2535–2548

Golovina E, Khloponina V, Tsiglianu P, Zhu R (2023) Organizational, economic and regulatory aspects of groundwater resources extraction by individuals (case of The Russian Federation). Resources 12(8):89

Gopalakrishnan C, Tortajada C, Biswas AK (2005) Water institutions: policies, performance and prospects. Springer, Berlin

Grafton RQ, Williams J, Perry CJ, Molle F, Ringler C, Steduto P, Udall B, Wheeler SA, Wang Y, Garrick D (2018) The paradox of irrigation efficiency. Science 361(6404):748–750

Grafton RQ, Garrick D, Manero A, Do TN (2019) The water governance reform framework: overview and applications to Australia, Mexico, Tanzania, USA and Vietnam. Water 11(1):137

Greitens TJ (2016) Assessing transboundary governance capacity in the Great Lakes areas of concern. Int J Water Gov 4:53–72

Grey D, Sadoff CW (2007) Sink or swim? Water security for growth and development. Water Policy 9(6):545–571

Grigg NS (2018) Collective action in multilevel water governance and management: variation by scale and problem type. Int J Water Gov 6:1–18

Grison C, Koop S, Eisenreich S, Hofman J, Chang I-S, Wu J, Savic D, Van Leeuwen K (2023) Integrated water resources management in cities in the world: global challenges. Water Resour Manag 37(6):2787–2803

Gupta J, Pahl-Wostl C (2013) Editorial on global water governance. Ecol Soc 18(4):54

Gutiérrez-Malaxechebarría A-M (2013) Informal irrigation in the Colombian Andes: local practices, national agendas, and options for innovation. Mt Res Dev 33(3):260–268

Handayani W, Dewi SP, Septiarani B (2023) Toward adaptive water governance: an examination on stakeholders engagement and interactions in Semarang City, Indonesia. Environ Dev Sustain 25(2):1914–1943

Hayden H, Tsvetanov T (2019) The effectiveness of urban irrigation day restrictions in California. Water Econ Policy 5(3):1–29

Heckman JJ, Smith J, Clements N (1997) Making the most out of programme evaluations and social experiments: accounting for heterogeneity in programme impacts. Rev Econ Stud 64(4):487–535

Heinmiller T (2016) Institutions and transboundary governance capacity in the Great Lakes Basin: the case of irrigation water-takings. Int J Water Gov 4:33–52

Hjorth P, Madani K (2023) Adaptive water management: on the need for using the post-WWII science in water governance. Water Resour Manag 37(6):2247–2270

Hoanh CT, Suhardiman D (2014) Irrigation development in the Vietnamese Mekong Delta: towards polycentric water governance? Int J Water Gov 2(2–3):61–82

Hoekstra AY (2011) The global dimension of water governance: why the river basin approach is no longer sufficient and why cooperative action at global level is needed. Water 3(1):21–46

Hoekstra AY, Chapagain AK, Van Oel PR (2019) Progress in water footprint assessment: towards collective action in water governance. Water 11:1070

Horne J, Grafton RQ (2019) The Australian water markets story: incremental transformation. In: Successful public policy. ANU Press, Canberra, p 165

Huang Q, Rozelle S, Wang J, Huang J (2009) Water management institutional reform: a representative look at northern China. Agric Water Manag 96(2):215–225

Huang Q, Wang J, Easter KW, Rozelle S (2010) Empirical assessment of water management institutions in northern China. Agric Water Manag 98(2):361–369

Hunt RC (2007) Communal irrigation: a comparative perspective. In: A world of water. Brill, Leiden, pp 185–208

Huntjens P, Lebel L, Pahl-Wostl C, Camkin J, Schulze R, Kranz N (2012) Institutional design propositions for the governance of adaptation to climate change in the water sector. Glob Environ Chang 22(1):67–81

Ibrahim IA, Farah PD (2025) Where climate meets water in transboundary systems: breaking boundaries through law. Water Int 50(3–4):329–349

Iglesias E, Garrido A, Gómez-Ramos A (2007) Economic drought management index to evaluate water institutions' performance under uncertainty. Aust J Agric Resour Econ 51(1):17–38

Ishaque W, Mukhtar M, Tanvir R (2023) Pakistan's water resource management: ensuring water security for sustainable development. Front Environ Sci 11:1096747

Islam S, Madani K (2017) Water diplomacy in action: contingent approaches to managing complex water problems, vol 1. Anthem Press, London

Iyiola AO, Kolawole AS, Ajayi FO, Ogidi OI, Ogwu MC (2024) Sustainable water use and management for agricultural transformation in Africa. In: Water resources management for rural development. Elsevier, Amsterdam, pp 287–300

Janssen MA, Anderies JM (2013) A multi-method approach to study robustness of social–ecological systems: the case of small-scale irrigation systems. J Inst Econ 9(4):427–447

Jibat E, Senbeta F, Zeleke T, Hagos F (2024) The role and interplay of institutions in water governance in the Central Rift Valley of Ethiopia. F1000Res 12:1434

Jiménez A, Saikia P, Giné R, Avello P, Leten J, Lymer BL, Schneider K, Ward R (2020) Unpacking water governance: a framework for practitioners. Water 12(3):827

Jogo W, Hassan R (2010) Balancing the use of wetlands for economic well-being and ecological security: the case of the Limpopo wetland in southern Africa. Ecol Econ 69(7):1569–1579

Jones-Crank JL (2024) A multi-case institutional analysis of water–energy–food nexus governance. Sustain Sci 19(4):1277–1291

Jyotishi A, Rout S (2005) Water rights in Deccan region: insights from Baliraja and other water institutions. Econ Polit Wkly 40:149–156

Kahil MT, Connor JD, Albiac J (2015) Efficient water management policies for irrigation adaptation to climate change in Southern Europe. Ecol Econ 120:226–233

Kajisa K, Dong B (2017) The effects of volumetric pricing policy on farmers' water management institutions and their water use: the case of water user organization in an irrigation system in Hubei, China. World Bank Econ Rev 31(1):220–240

Kallis G (2010) Coevolution in water resource development: the vicious cycle of water supply and demand in Athens, Greece. Ecol Econ 69(4):796–809

Kayaga S, Mugabi J, Kingdom W (2013) Evaluating the institutional sustainability of an urban water utility: a conceptual framework and research directions. Util Policy 27:15–27

Keen M (2003) Integrated water management in the South Pacific: policy, institutional and socio-cultural dimensions. Water Policy 5(2):147–164

Khan HF, Morzuch BJ, Brown CM (2017) Water and growth: an econometric analysis of climate and policy impacts. Water Resour Res 53(6):5124–5136

Kibaroglu A (2020) The role of irrigation associations and privatization policies in irrigation management in Turkey. Water Int 45(2):83–90

Kibaroglu A (2025) Historical review of formal and informal water institutions in the Euphrates-Tigris Region with a specific focus on water relations between Turkey and Iraq. In: World scientific handbook of transboundary water management: science, economics, policy and politics: volume 3: the role of formal and informal institutions in managing transboundary basins. World Scientific, Singapore, pp 199–222

Kimmich C, Villamayor-Tomas S (2019) Assessing action situation networks: a configurational perspective on water and energy governance in irrigation systems. Water Econ Policy 5(1):1850005

Kiparsky M, Milman A, Owen D, Fisher AT (2017) The importance of institutional design for distributed local-level governance of groundwater: the case of California's Sustainable Groundwater Management Act. Water 9(10):755

Kirchhoff CJ, Dilling L (2016) The role of US states in facilitating effective water governance under stress and change. Water Resour Res 52(4):2951–2964

Klassert C, Yoon J, Sigel K, Klauer B, Talozi S, Lachaut T, Selby P, Knox S, Avisse N, Tilmant A (2023) Unexpected growth of an illegal water market. Nat Sustain 6(11):1406–1417

Kolinjivadi V, Adamowski J, Kosoy N (2014) Recasting payments for ecosystem services (PES) in water resource management: a novel institutional approach. Ecosyst Serv 10:144–154

Kremer M, Leino J, Miguel E, Zwane AP (2011) Spring cleaning: rural water impacts, valuation, and property rights institutions. Q J Econ 126(1):145–205

Krishnaraj M (2011) Women and water: issues of gender, caste, class and institutions. Econ Polit Wkly 46:37–39

Kuhn A, Britz W, Willy DK, van Oel P (2016) Simulating the viability of water institutions under volatile rainfall conditions—the case of the Lake Naivasha Basin. Environ Model Softw 75:373–387

Kumar MD (2018) Water policy science and politics: an Indian perspective. Elsevier, Amsterdam

Kuwayama Y, Brozović N (2013) The regulation of a spatially heterogeneous externality: tradable groundwater permits to protect streams. J Environ Econ Manag 66(2):364–382

Lai CH, Zhao J (2023) Decomposing the institutional structure of Interstate River basins using institutional grammar and institutional analysis and development framework. J Water Resour Plan Manag 149(7):04023022

Lam WF (2006) Foundations of a robust social-ecological system: irrigation institutions in Taiwan. J Inst Econ 2(2):203–226

Lawless KL, Garcia M, White DD (2024) Institutional analysis of water governance in the Colorado River Basin, 1922–2022. Front Water 6:1451854

Layzer JA, Schulman A (2013) IWRM in the United States: integration in the Chesapeake Bay program. Int J Water Gov 1(3–4):237–264

Lefebvre M, Gangadharan L, Thoyer S (2012) Do security-differentiated water rights improve the performance of water markets? Am J Agric Econ 94(5):1113–1135

Li H, Zhao J (2018) Rebound effects of new irrigation technologies: the role of water rights. Am J Agric Econ 100(3):786–808

Li W, Beresford M, Song G (2011) Market failure or governmental failure? A study of China's water abstraction policies. China Q 208:951–969

Li M, Xu W, Rosegrant MW (2017) Irrigation, risk aversion, and water right priority under water supply uncertainty. Water Resour Res 53(9):7885–7903

Lubell M, Edelenbos J (2013) Integrated water resources management: a comparative laboratory for water governance. Int J Water Gov 1(3–4):177–196

Ludwig F, Kabat P, van Schaik H, van der Valk M (2012) Climate change adaptation in the water sector. Routledge, Milton Park

Madani K, Dinar A (2012a) Cooperative institutions for sustainable common pool resource management: application to groundwater. Water Resour Res 48(9)

Madani K, Dinar A (2012b) Non-cooperative institutions for sustainable common pool resource management: application to groundwater. Ecol Econ 74:34–45

Madani K, Dinar A (2013) Exogenous regulatory institutions for sustainable common pool resource management: application to groundwater. Water Resour Econ 2:57–76

Makate C, Wang R, Tatsvarei S (2018) Water footprint concept and methodology for warranting sustainability in human-induced water use and governance. Sustain Water Resour Manag 4(1):91–103

Mao Z, Xue X, Tian H, Michael AU (2019) How will China realize SDG 14 by 2030?—A case study of an institutional approach to achieve proper control of coastal water pollution. J Environ Manag 230:53–62

Märker C, Venghaus S, Hake J-F (2018) Integrated governance for the food–energy–water nexus—the scope of action for institutional change. Renew Sust Energ Rev 97:290–300

Marshall G, Connell D, Taylor BM (2013) Australia's Murray-Darling basin: a century of polycentric experiments in cross-border integration of water resources management. Int J Water Gov 1:231–251

Marshall D, Salamé L, Wolf AT (2017) A call for capacity development for improved water diplomacy. In: Water diplomacy in action: contingent approaches to managing complex water problems, vol 1. Anthem Press, London, p 141

Martín Velasco MJ, Calderon G, Lima ML, Matencón CL, Massone HE (2023) Water governance challenges at a local level: implementation of the OECD water governance indicator framework in the General Pueyrredon Municipality, Buenos Aires province, Argentina. Water Policy 25(7):623–638

Martínez-Valderrama J, Olcina J, Delacámara G, Guirado E, Maestre FT (2023) Complex policy mixes are needed to cope with agricultural water demands under climate change. Water Resour Manag 37(6):2805–2834

Means EG III, Brueck T, Manning A, Dixon L (2002) The coming crisis: water institutions and infrastructure. Am Water Works Assoc J 94(1):34

Meinzen-Dick R (2007) Beyond panaceas in water institutions. Proc Natl Acad Sci 104(39):15200–15205

Meissner R, Funke N, Nienaber S, Ntombela C (2013) The status quo of research on South Africa's water resource management institutions. Water SA 39(5):721–732

Ménard C (2017) Meso-institutions: the variety of regulatory arrangements in the water sector. Util Policy 49:6–19

Michalak D (2020) Adapting to climate change and effective water management in Polish agriculture—at the level of government institutions and farms. Ecohydrol Hydrobiol 20(1):134–141

Middleton C, Dore J (2015) Transboundary water and electricity governance in mainland Southeast Asia: linkages, disjunctures and implications. Int J Water Gov 3(1):93–120

Mirzaei A, Theesfeld I (2025) Diving deep into legal layers: institutional grammar's insight into Iran's groundwater laws. Water Policy 27(4):477–500

Mogomotsi PK, Mogomotsi GEJ, Matlhola DM (2018) A review of formal institutions affecting water supply and access in Botswana. Phys Chem Earth Parts A/B/C 105:283–289

Möllenkamp S, Kastens B (2012) Institutional adaptation to climate change: current status and future strategies in the Elbe Basin, Germany. Routledge, Milton Park

Moss T (2004) The governance of land use in river basins: prospects for overcoming problems of institutional interplay with the EU Water Framework Directive. Land Use Policy 21(1):85–94

Mottaleb KA, Krupnik TJ, Keil A, Erenstein O (2019) Understanding clients, providers and the institutional dimensions of irrigation services in developing countries: a study of water markets in Bangladesh. Agric Water Manag 222:242–253

Msuya TS, Lalika MCS (2018) Linking ecohydrology and integrated water resources management: institutional challenges for water management in the Pangani Basin, Tanzania. Ecohydrol Hydrobiol 18(2):174–191

Muchapondwa E, Stage J, Mungatana E, Kumar P (2018) Lessons from applying market-based incentives in watershed management. Water Econ Policy 4(3):1850011

Mujtaba G, Shah MUH, Hai A, Daud M, Hayat M (2024) A holistic approach to embracing the United Nation's Sustainable Development Goal (SDG-6) towards water security in Pakistan. J Water Process Eng 57:104691

Mukherji A (2007) Implications of alternative institutional arrangements in groundwater sharing: evidence from West Bengal. Econ Polit Wkly 42:2543–2564

Mukhtarov F, Fox S, Mukhamedova N, Wegerich K (2015) Interactive institutional design and contextual relevance: water user groups in Turkey, Azerbaijan and Uzbekistan. Environ Sci Policy 53:206–214

Muller H (2014) The South African experience on legal, institutional and operational aspects of the rights to water and sanitation. Aquat Procedia 2:35–41

Mulroy P (2017) The water problem: climate change and water policy in the United States. Brookings Institution Press, Washington, DC

Mwakalila E, Muneeni JM (2025) Role of local water institutions in water resources management in Sanya-Kware Sub-catchment of the Pangani Basin, Tanzania. J Basic Appl Res Int 31(1):1–18

Nauges C, Whittington D (2019) Social norms information treatments in the municipal water supply sector: some new insights on benefits and costs. Water Econ Policy 5(3):1850026

Nüsser M, Dame J, Parveen S, Kraus B, Baghel R, Schmidt S (2019) Cryosphere-fed irrigation networks in the northwestern Himalaya: precarious livelihoods and adaptation strategies under the impact of climate change. Mt Res Dev 39(2):R1–R11

Nyagumbo I, Rurinda J (2012) An appraisal of policies and institutional frameworks impacting on smallholder agricultural water management in Zimbabwe. Phys Chem Earth Parts A/B/C 47:21–32

O'Donnell E (2023) Water sovereignty for indigenous peoples: pathways to pluralist, legitimate and sustainable water laws in settler colonial states. PLoS Water 2(11):e0000144

Oberlack C, Eisenack K (2018) Archetypical barriers to adapting water governance in river basins to climate change. J Inst Econ 14(3):527–555

Olen B, Wu JJ, Langpap C (2016) Irrigation decisions for major west coast crops: water scarcity and climatic determinants. Am J Agric Econ 98(1):254–275

Olivier T, Vallury S (2024) Institutional fit and policy design in water governance: Nebraska's Natural Resources Districts. Policy Stud J 52(4):809–832

Ouyang R, Mu E, Yu Y, Chen Y, Hu J, Tong H, Cheng Z (2024) Assessing the effectiveness and function of the water resources tax policy pilot in China. Environ Dev Sustain 26(1):2637–2653

Özerol G (2013) Institutions of farmer participation and environmental sustainability: a multi-level analysis from irrigation management in Harran Plain, Turkey. Int J Commons 7(1):73–91

Özerol G, Kruijf JV-d, Brisbois MC, Flores CC, Deekshit P, Girard C, Knieper C, Mirnezami SJ, Ortega-Reig M, Ranjan P (2018) Comparative studies of water governance. Ecol Soc 23(4)

Pahl-Wostl C, Knieper C (2023) Pathways towards improved water governance: the role of polycentric governance systems and vertical and horizontal coordination. Environ Sci Policy 144:151–161

Pahl-Wostl C, Lukat E, Stein U, Troeltzsch J, Yousefi A (2023) Improving the socio-ecological fit in water governance by enhancing coordination of ecosystem services used. Environ Sci Policy 139:11–21

Peat M, Moon K, Dyer F, Johnson W, Nichols SJ (2017) Creating institutional flexibility for adaptive water management: insights from two management agencies. J Environ Manag 202:188–197

Plumb ST, Paveglio T, Jones KW, Miller BA, Becker DR (2018) Differentiated reactions to payment for ecosystem service programs in the Columbia River Basin: a qualitative study exploring irrigation district characteristics as local common-pool resource management institutions in Oregon, USA. Int J Commons 12(1):202–224

Polonenko LMM, Hamouda MA, Mohamed MM (2020) Essential components of institutional and social indicators in assessing the sustainability and resilience of urban water systems: challenges and opportunities. Sci Total Environ 708:135159

Posthumus H, Rouquette JR, Morris J, Gowing DJG, Hess TM (2010) A framework for the assessment of ecosystem goods and services; a case study on lowland floodplains in England. Ecol Econ 69(7):1510–1523

Pradhan D, Ranjan R (2015) Do institutional programs aimed at groundwater augmentation affect crop choice decisions under groundwater irrigation? Empirical evidence from Andhra Pradesh, India. Water Econ Policy 1(2):1550002

Prniyazova A, Turaeva S, Turgunov D, Jarihani B (2025) Sustainable transboundary water governance in Central Asia: challenges, conflicts, and regional cooperation. Sustainability 17(11):4968

Programme, World Water Assessment (2009) The United Nations world water development report. UNESCO Publications, Paris

Pulido-Velazquez M, Ward FA (2017) Comparison of water management institutions and approaches in the United States and Europe—what can we learn from each other? In: Competition for water resources. Elsevier, Amsterdam, pp 423–441

Punjabi B, Johnson CA (2019) The politics of rural–urban water conflict in India: untapping the power of institutional reform. World Dev 120:182–192

Rai RK, Nepal M, Bhatta LD, Das S, Khadayat MS, Somanathan E, Baral K (2019) Ensuring water availability to water users through incentive payment for ecosystem services scheme: a case study in a small hilly town of Nepal. Water Econ Policy 5(4):1850002

Rak G, Zec D, Markovčić Kostelac M, Joksimović D, Gollasch S, David M (2019) The implementation of the ballast water management convention in the Adriatic Sea through states' cooperation: the contribution of environmental law and institutions. Mar Pollut Bull 147:245–253

Regnacq C, Dinar A, Hanak E (2016) The gravity of water: water trade frictions in California. Am J Agric Econ 98(5):1273–1294

Rieu-Clarke A (2015) Transboundary hydropower projects seen through the lens of three international legal regimes: foreign investment, environmental protection and human rights. Int J Water Gov 3(1):27–48

Rogers BC, Brown RR, De Haan FJ, Deletic A (2015) Analysis of institutional work on innovation trajectories in water infrastructure systems of Melbourne, Australia. Environ Innov Soc Trans 15:42–64

Roßner R, Zikos D (2018) The role of homogeneity and heterogeneity among resource users on water governance: lessons learnt from an economic field experiment on irrigation in Uzbekistan. Water Econ Policy 4(3):1850008

Ruiters C, Matji MP (2015) Water institutions and governance models for the funding, financing and management of water infrastructure in South Africa. Water SA 41(5):660–676

Rusca M, Schwartz K (2014) 'Going with the grain': accommodating local institutions in water governance. Curr Opin Environ Sustain 11:34–38

Sakketa TG (2018) Institutional bricolage as a new perspective to analyse institutions of communal irrigation: implications towards meeting the water needs of the poor communities. World Dev Perspect 9:1–11

Saleth RM (2011) Water scarcity and climatic change in India: the need for water demand and supply management. Hydrol Sci J 56(4):671–686

Saleth RM (2018) The institutional economics of water. Water Econ Policy 4(3):1802003. (11 pages)

Saleth RM, Dinar A (2004) The institutional economics of water: a cross-country analysis of institutions and performance. World Bank Publications, Washington, DC

Saleth RM, Bassi N, Kumar MD (2016) Role of institutions, infrastructures, and technologies in meeting global agricultural water challenge. Choices 31(3):1–7

Santos E, Carvalho M, Martins S (2023) Sustainable water management: understanding the socioeconomic and cultural dimensions. Sustainability 15(17):13074

Saravanan VS (2015) Agents of institutional change: the contribution of new institutionalism in understanding water governance in India. Environ Sci Policy 53:225–235

Satoh E (2019) Water demand fluctuations, non-transferable water rights, and technical inefficiency in Japan's water sector. Water Econ Policy 5(3):1850028

Schmeier S, Blumstein S (2025) The role of basin water treaties and basin organizations in managing transboundary water resources: taking stock of current practice. In: World scientific handbook of transboundary water management: science, economics, policy and politics: volume 3: the role of formal and informal institutions in managing transboundary basins. World Scientific, Singapore, pp 51–77

Schnegg M, Bollig M (2016) Institutions put to the test: community-based water management in Namibia during a drought. J Arid Environ 124:62–71

Schnegg M, Linke T (2015) Living institutions: sharing and sanctioning water among pastoralists in Namibia. World Dev 68:205–214

Scott CA, Pablos NP (2011) Innovating resource regimes: water, wastewater, and the institutional dynamics of urban hydraulic reach in Northwest Mexico. Geoforum 42(4):439–450

Scott CA, Pierce SA, Pasqualetti MJ, Jones AL, Montz BE, Hoover JH (2011) Policy and institutional dimensions of the water–energy nexus. Energy Policy 39(10):6622–6630

Sears L, Lim D, Lawell C-YCL (2018) The economics of agricultural groundwater management institutions: the case of California. Water Econ Policy 4(3):1850003

Sears L, Lim D, Lawell C-YCL (2019) Spatial groundwater management: a dynamic game framework and application to California. Water Econ Policy 5(1):1850019

Sehring J, Sharipova B, Assubayeva A (2024) The politics of water governance in Central Asia: institutionalizing river basin management. In: Handbook on the governance and politics of water resources. Edward Elgar Publishing, Cheltenham, pp 205–217

Seijger C, Hellegers P (2023) How do societies reform their agricultural water management towards new priorities for water, agriculture, and the environment? Agric Water Manag 277:108104

Shah SH, Narain V (2019) Re-framing India's "water crisis": an institutions and entitlements perspective. Geoforum 101:76–79

Shah T, van Koppen B (2006) Is India ripe for integrated water resources management? Fitting water policy to national development context. Econ Polit Wkly 41:3413–3421

Shivakoti GP, Bastakoti RC (2006) The robustness of Montane irrigation systems of Thailand in a dynamic human–water resources interface. J Inst Econ 2(2):227–247

Sifundza LS, van der Zaag P, Masih I (2019) Evaluation of the responses of institutions and actors to the 2015/2016 El Niño drought in the Komati catchment in Southern Africa: lessons to support future drought management. Water SA 45(4):547–559

Sithirith M, Sok S, De Silva S (2023) Legal and institutional analysis of water resource management and development in Cambodia. WorldFish, Penang

Smajgl A, Heckbert S, Ward J, Straton A (2009) Simulating impacts of water trading in an institutional perspective. Environ Model Softw 24(2):191–201

Söderbaum F (2015) Rethinking the politics of transboundary water management: the case of the Zambezi river basin. Int J Water Gov 3(3):1–12

Söderberg C (2016) Complex governance structures and incoherent policies: implementing the EU water framework directive in Sweden. J Environ Manag 183:90–97

Sokile CS, Kashaigili JJ, Kadigi RMJ (2003) Towards an integrated water resource management in Tanzania: the role of appropriate institutional framework in Rufiji Basin. Phys Chem Earth Parts A/B/C 28(20–27):1015–1023

Sosa M, Zwarteveen M (2014) The institutional regulation of the sustainability of water resources within mining contexts: accountability and plurality. Curr Opin Environ Sustain 11:19–25

Sowby RB, South AJ (2023) Innovative water rates as a policy tool for drought response: two case studies from Utah, USA. Util Policy 82:101570

Stoa RB (2017) Water governance in Haiti: an assessment of laws and institutional capacities. Tulane Environ Law J 29(2):243–286

Suarez Bosa M (2015) Water institutions and management in Cape Verde. Water 7(6):2641–2655

Suleymanov F (2024) The institutionalization of the Kura-Aras River Basin for effective management of water resources. Int J River Basin Manag 23:675–685

Syed T, Choudhury E (2018) Scale interactions in transboundary water governance of Indus river. Int J Water Gov 6:64–84

Taylor BM, Eberhard R (2020) Practice change, participation and policy settings: a review of social and institutional conditions influencing water quality outcomes in the Great Barrier Reef. Ocean Coast Manag 190:105156

Thiel A (2014) Developing an analytical framework for reconstructing the scalar reorganization of water governance as institutional change: the case of Southern Spain. Ecol Econ 107:378–391

Thomas A (1995) Regulating pollution under asymmetric information: the case of industrial wastewater treatment. J Environ Econ Manag 28(3):357–373

Thomas V, Warner J (2014) River Basin multi-stakeholder platforms: the practice of 'good water governance' in Afghanistan. Int J Water Gov 2(2–3):105–132

Tiger MW, Eskaf S, Hughes JA (2011) Implications of residential irrigation metering for customers' expenditures and demand. J Am Water Works Assoc 103(12):30–41

Tingey-Holyoak J (2014) Sustainable water storage by agricultural businesses: strategic responses to institutional pressures. J Bus Res 67(12):2590–2602

Tir J, Stinnett DM (2012) Weathering climate change: can institutions mitigate international water conflict? J Peace Res 49(1):211–225

Tortajada C, Joshi YK (2013) Water resources management and governance as part of an overall framework for growth and development. Int J Water Gov 1(3–4):285–306

Turgul A, Porta EL, McCracken M, Wolf AT (2025) Changing values: trends in international freshwater agreements. In: World scientific handbook of transboundary water management: science, economics, policy and politics: volume 3: the role of formal and informal institutions in managing transboundary basins. World Scientific, Singapore, pp 19–50

Ulibarri N, Scott TA (2019) Environmental hazards, rigid institutions, and transformative change: how drought affects the consideration of water and climate impacts in infrastructure management. Glob Environ Chang 59:102005

Van De Meene SJ, Brown RR (2009) Delving into the "institutional black box": revealing the attributes of sustainable urban water management regimes. JAWRA J Am Water Resour Assoc 45(6):1448–1464

Van den Hurk M, Mastenbroek E, Meijerink S (2014) Water safety and spatial development: an institutional comparison between the United Kingdom and The Netherlands. Land Use Policy 36:416–426

Van der Kooij S, Zwarteveen M, Kuper M (2015) The material of the social: the mutual shaping of institutions by irrigation technology and society in Seguia Khrichfa, Morocco. Int J Commons 9(1)

VanNijnatten D, Johns C, Friedman KB, Krantzberg G (2016) Assessing adaptive transboundary governance capacity in the Great Lakes Basin: the role of institutions and networks. Int J Water Gov 4:7–32

Varady RG, Albrecht TR, Modak S, Wilder MO, Gerlak AK (2023) Transboundary water governance scholarship: a critical review. Environments 10(2):27

Venot J-P, Suhardiman D (2014) Governing the ungovernable: practices and circumstances of governance in the irrigation sector. Int J Water Gov 2(2–3):41–60

Verweij S, Van Meerkerk I, Koppenjan J, Geerlings H (2014) Institutional interventions in complex urban systems. Emerg Complex Organ 16(1):7–23

Videira N, Antunes P, Santos R (2009) Scoping river basin management issues with participatory modelling: the Baixo Guadiana experience. Ecol Econ 68(4):965–978

Villamayor-Tomas S (2017) The Water–Energy Nexus in Europe and Spain: an institutional analysis from the perspective of the Spanish irrigation sector. In: Competition for water resources. Elsevier, Amsterdam, pp 105–122

Villamayor-Tomas S (2018) Disturbance features, coordination and cooperation: an institutional economics analysis of adaptations in the Spanish irrigation sector. J Inst Econ 14(3):501–526

Vitale C, Meijerink S, Moccia FD (2023) Urban flood resilience, a multi-level institutional analysis of planning practices in the Metropolitan City of Naples. J Environ Plan Manag 66(4):813–835

Wan Z, Chen J, Sperling D (2018) Institutional barriers to the development of a comprehensive ballast-water management scheme in China: perspective from a multi-stream policy model. Mar Policy 91:142–149

Wang X, Otto IM, Yu L (2013) How physical and social factors affect village-level irrigation: an institutional analysis of water governance in northern China. Agric Water Manag 119:10–18

Washington-Ottombre C, Evans TP (2019) Relationships between institutional success and length of tenure in a Kenyan irrigation scheme. Int J Commons 13(1):329–352

Water C (2019) Leaving no one behind. The United Nations world water development report

Wegerich K, Warner J, Tortajada C (2014) Water sector governance: a return ticket to anarchy. Int J Water Gov 2(2–3):7–20

Wheeler SA, Owens K, Zuo A (2024) Is there public desire for a federal takeover of water resource management in Australia? Water Res 248:120861

Wutich A, York AM, Brewis A, Stotts R, Roberts CM (2012) Shared cultural norms for justice in water institutions: results from Fiji, Ecuador, Paraguay, New Zealand, and the US. J Environ Manag 113:370–376

Wutich A, Budds J, Eichelberger L, Geere J, Harris LM, Horney JA, Jepson W, Norman E, O'Reilly K, Pearson AL (2017) Advancing methods for research on household water insecurity: studying entitlements and capabilities, socio-cultural dynamics, and political processes, institutions and governance. Water Secur 2:1–10

Wyborn C, Van Kerkhoff L, Colloff M, Alexandra J, Olsson R (2023) The politics of adaptive governance: water reform, climate change, and First Nations' justice in Australia's Murray-Darling Basin. Ecol Soc 28:4

Xie Y, Zilberman D (2016) Theoretical implications of institutional, environmental, and technological changes for capacity choices of water projects. Water Resour Econ 13:19–29

Yang H, Zhang X, Zehnder AJB (2003) Water scarcity, pricing mechanism and institutional reform in northern China irrigated agriculture. Agric Water Manag 61(2):143–161

Zhang C-Y, Oki T (2023) Water pricing reform for sustainable water resources management in China's agricultural sector. Agric Water Manag 275:108045

Zhang G, Hoekstra AY, Mathews RE (2013) Water footprint assessment (WFA) for better water governance and sustainable development, editorial. Water Resour Ind 1:1–6

Zhang L, Zhu X, Heerink N, Shi X (2014) Does output market development affect irrigation water institutions? Insights from a case study in northern China. Agric Water Manag 131:70–78

Zikos D, Roggero M (2013) The patronage of thirst: exploring institutional fit on a divided Cyprus. Ecol Soc 18(2)

Zinabu E, Alamirew T, Whitehead P, Charles K, Abebe Y, Zeleke G (2024) Evaluating the structures and arrangements of water institutions to include in-stream modeling for water quality management and control pollution: insights from the Awash Basin, Ethiopia. World Water Policy 10(1):233–243

Zuo A, Wheeler SA, Adamowicz WL, Boxall PC, Hatton-MacDonald D (2016) Measuring price elasticities of demand and supply of water entitlements based on stated and revealed preference data. Am J Agric Econ 98(1):314–332

MIX
Papier aus verantwortungsvollen Quellen
Paper from responsible sources
FSC® C105338

If you have any concerns about our products,
you can contact us on
ProductSafety@springernature.com

In case Publisher is established outside the EU,
the EU authorized representative is:
Springer Nature Customer Service Center GmbH
Europaplatz 3, 69115 Heidelberg, Germany

Printed by Libri Plureos GmbH
in Hamburg, Germany